Fundamentals of Electro-Optic Systems Design

Communications, Lidar, and Imaging

Using fundamentals of communication theory, thermodynamics, information theory and propagation theory, this book explains the universal principles underlying a diverse range of electro-optical systems. From visible / infra-red imaging, to free space optical communications and laser remote sensing, the authors relate key concepts in science and device engineering to practical systems issues.

A broad spectrum of coherent and incoherent electro-optical systems is considered, accompanied by many real-world examples. The authors also present new insights into the fundamental limitations of these systems when operating through deleterious channels.

Accompanied by online examples of processed images and ideas, this uniquely tailored guide to the fundamental principles underlying modern electro-optical systems is an essential reference for all practicing engineers, graduate students, and academic researchers in optical engineering.

Sherman Karp received his Ph.D. from the University of Southern California, and has gone on to work with NASA, and as Principal Scientist at DARPA. The author of several books, he has also been awarded the SECDEF Medal for Meritorious Civilian Service, and NOSC "Scientist of the Year." He is a Fellow of the IEEE.

Larry B. Stotts received his Ph.D. from the University of California at San Diego, and is a former Deputy Office Director for the Strategic Technology Office, DARPA. He has been awarded two DARPA Technical Achievement Awards, two SECDEF Medals for Meritorious Civilian Service, and the NOSC Technical Director's Award. He is a Fellow of the IEEE and SPIE.

"With the completion of *Fundamentals of Electro-Optic Systems Design*, Sherman Karp and Larry B. Stotts have created a single comprehensive book for anyone having anything to do with the vast field of electro-optics. The detailed systems design principles, examples, charts, graphs, and methods bring together under one cover the information to handle the applications given by the subtitle *Communications, Lidar, and Imaging*. The basic theories and their relationship to real-world hardware constraints such as noise and scattering are covered in full detail with necessary citations to decades of electro-optics research. From a systems design point-of-view, Karp and Stotts blend Lidar, laser communications, and imaging into a logical path to analyze, design, and test complex electro-optics. The communication chapters covering modulation, coding, and propagation in various media are not found anywhere else unless one wades through thousands of research papers and reports. If you are a scientist or engineer who has to manipulate photons, *Fundamentals of Electro-Optic Systems Design* belongs on your bookshelf – near the front."

Robert K. Tyson, The University of North Carolina at Charlotte

"This book uniquely treats electro-optical system design from an engineering viewpoint emphasizing real world applications and where theory works and does not work. These perspectives make this book a must-have reference for the scientist or engineer involved with electro-optical system design."

Tony Tether, Former DARPA Director 2001 to 2009

"*Fundamentals of Electro-Optic Systems Design* is a comprehensive and authoritative treatment of free-space optical communications and Lidar. Topics range from diffraction, photoelectric detection, effects of scattering and optical turbulence, and even signal coding, modulation and error correction."

Joseph W. Goodman, Stanford University

"The book is written by very knowledgeable and very experienced individuals in the field of electro-optical systems. Their writing and explanations make the material very accessible. It is clear and well presented."

Ronald Phillips, University of Central Florida

Fundamentals of Electro-Optic Systems Design

Communications, Lidar, and Imaging

SHERMAN KARP
LARRY B. STOTTS

CAMBRIDGE
UNIVERSITY PRESS

CAMBRIDGE
UNIVERSITY PRESS

University Printing House, Cambridge CB2 8BS, United Kingdom

One Liberty Plaza, 20th Floor, New York, NY 10006, USA

477 Williamstown Road, Port Melbourne, VIC 3207, Australia

314-321, 3rd Floor, Plot 3, Splendor Forum, Jasola District Centre, New Delhi - 110025, India

79 Anson Road, #06-04/06, Singapore 079906

Cambridge University Press is part of the University of Cambridge.

It furthers the University's mission by disseminating knowledge in the pursuit of
education, learning and research at the highest international levels of excellence.

www.cambridge.org
Information on this title: www.cambridge.org/9781107021396

First published 2013

A catalogue record for this publication is available from the British Library

Library of Congress Cataloging in Publication data
Karp, Sherman.
Fundamentals of electro-optic systems design : communications, lidar, and imaging / Sherman
Karp, Larry B. Stotts.
    p.   cm.
ISBN 978-1-107-02139-6 (hardback)
1. Optical communications.   2. Optical radar.   3. Imaging systems.   I. Stotts, Larry B.
II. Title.
TK5103.59.K337   2012
621.382'7–dc23   2012021566

ISBN 978-1-107-02139-6 Hardback

Additional resources for this publication at www.cambridge.org/karp-stotts

In Memorandum

The authors dedicate this book to our esteemed friend and colleague, Professor Robert M. Gagliardi, who recently passed away. We have known Bob for almost fifty years, and we will miss him. He was truly a great and gentle man.

Contents

Preface

The authors have been active participants in the area of electro-optic systems for over four decades, covering the introduction of laser systems and low loss optical fibers and the institutionalizing of photonic systems into everyday life. Yet for all the literature that exists, and all the work that has been accomplished, we felt that no single book existed that integrated the entire field of electro-optics, reaching back to all the fundamental building blocks and providing enough examples to be useful to practicing engineers. After much discussion and a slow start, we decided first to reference as much material as possible, bringing forth only the highlights necessary to guide researchers in the field. Then we decided to minimize mathematical developments by relegating them, as much as possible, to explanatory examples. What has evolved in our development is a clear statement of the duality of time and space in electro-optic systems. This had been touched upon in our earlier work, but has been brought forth clearly in this book in the duality of modulation index in time, and contrast in space. In doing so, and in other areas, we feel that this book contains new material with regard to the processing of spatial images which have propagated through deleterious channels. We feel that this book contains much new material in the areas of communications and imaging through deleterious channels.

In Chapter 1, we reach back to the true foundations of modern physics, the establishment of the first two laws of thermodynamics. While taken for granted, it is the first law that explains why we can see stars at the edge of the universe, and governs the radiant properties of propagating systems. The second law and the insight of Claude Shannon have created the modern field of Information Theory. Using his fundamental definitions of channel capacity we are able to establish the duality of time and space in electro-optics. This requires one basic mathematical development that is included in Appendix A, and is developed in Chapters 3 and 4.

In Chapter 2, we present the development of Maxwell's equations, which we correlate with radiant properties of propagating fields. We bring this development to the point where we can establish the conditions under which duality applies. In the process we demonstrate how the radiant properties can be used to make engineering calculations in both temporal and spatial systems, passing from the transmitting optics through to the detector.

In Chapter 3, we focus on the behavior of the photo-detection system. Using Appendix A, we introduce the duality between temporal systems and spatial systems. We identify the information-bearing portion of the received signal (as defined by Shannon) and the modulation index of incoherent communication systems.

In Chapter 4, we delve more deeply into spatial systems and the effects on imaging. Again using Appendix A, we show the duality between the modulation index in temporal systems and contrast in spatial systems. This is related to the classic definition and merely extends it based upon the newer understanding of the behavior of electro-optic systems. We then proceed further in digital communication systems and show how we develop optimum systems under a few different constraints.

In Chapter 5, we investigate in detail the concepts of contrast and visibility and their relationship to signal-to-noise ratio. We then discuss the duality of modulation index in communication theory and contrast in imaging. We also introduce the concept of sub-visibility imaging, which allows a direct trade-off between improved contrast and deteriorated signal-to-noise ratio. Imagery demonstrating this trade-off is presented, and applications are discussed.

In Chapter 6, we review various optical modulation schemes used in optical communications. This will include a discussion of spectral efficiency and the energy per bit per unilateral noise density, parameters used extensively in RF communications, but just coming into use in optical communications. These latter aspects help us determine the optimal means for optical communications, especially when erbium-doped fiber amplifiers are involved.

In Chapter 7, we provide the basics on block forward error correction (FEC) encodings, and highlight the arguably most popular FEC, Reed-Solomon codes. We then will note other important types of FEC schemes that can be employed, based on the authors' bias. This material is not meant to be an exhaustive survey of FEC coding, but rather included in this book to give the reader a basic background and knowledge of FEC to illustrate their utility in optical communications today. Application of FEC to optical systems will be discussed in a subsequent chapter.

In Chapter 8, we discuss some of the key aspects of the signal modulation and coding schemes used in fiber optic and free-space optical communications systems today. Most notably, we will review the use of return-to-zero and non-return-to-zero in coding the information streams and see their effect on systems performance, as well as receiver sensitivity.

In Chapter 9, we lay out the fundamentals of lidar and address a variety of applications. Although each of these applications has a huge area of research behind it, there are nevertheless some common threads which we discuss in the context of our development. We also reference some relevant papers in these areas for those interested in further investigation.

In Chapter 10, we discuss the performance of incoherent and coherent communications systems when operating in the optical turbulence channel. We will discuss a new statistical link budget approach for characterizing incoherent FSOC link performance, and compare experimental results with statistical predictions.

In Chapter 11, we discuss a set of approaches for optical communications in diffusive scattering channel, augmenting the results described in Chapter 9. Specifically, we highlight the accepted models for characterizing the statistical channel effects and techniques for providing communications in this channel.

As with many fields, we stand on the shoulders of great men and women whose contributions have shaped the field of engineering, especially optical system engineering. Of those researchers, the authors would like to acknowledge five pillars that have influenced the authors greatly through their teachings, personal interactions and insightful contributions – Professors Robert S. Kennedy, Irving S. Reed, Robert M. Gagliardi, Adolf Lohmann and Siebert Q. Duntley.

We also would like to thank our many colleagues who helped us over the years succeed in our optical systems projects to gain the insights we discuss in this book. They also have influenced us greatly. In writing this book, we want to acknowledge Drs. Larry Andrews, Juan Juarez, David Fried, Paul Kolodzy, Gary Lee, Ron Phillips, H. Alan Pike, David Young, and Mr. David Buck, Bob James, Todd Martin and Ned Plasson for their help and assistance.

In addition, we would like to acknowledge Dr. Anthony J. Tether, who over the years has encouraged and supported our various efforts in optical systems.

Finally, we recognize the biggest contributors to our book, our families, who have shown their patience during the preparation of this manuscript. Anyone who has imbedded themselves in an all-consuming goal knows the importance of the understanding and patience of loved ones.

Notation

$E[f(t)], \bar{f}$ expected value of $f(t)$

$m^2(t), \bar{m}^2$ modulation index; average value

$n(\vec{r}, t), \bar{n}$ Poissonrate parameter, average value

$2e^2\bar{n}B, (2ei_{dc}B)$ referred to as shot noise

PWE probability of symbol (word) error

(x, y) coordinates of aperture plane

(x_0, y_0) coordinates of receiver plane

$1/2B$ pulse width τ

a absorption coefficient radius of the aperture, length of the aperture

A area, optical amplitude

AO   adaptive optics

b volume scattering coefficient, length of the aperture

B bandwidth

$B(\mathbf{u})$ source irradiance function

BER   bit error rate

c, c_0 volume extinction coefficient; speed of light in free space

$C(\alpha), S(\alpha)$ Cornu spirals

C, C_0 information capacity (bits/s), constant, contrast; inherent contrast

C_{corr} correlation length

$C_n{}^2$ refractive index structure parameter

C_w curvature of the wave front

CNR   carrier-to-noise ratio

cx, cy centroid in x- and y-direction

d linear dimension of the photo-detector, aperture diameter

d_T, d_R transmitter, receiver aperture diameter

d_S beam diameter

d_d photo-detector diameter

d_R converging lens diameter

d_D diffraction-limited focus diameter

$D; D_s$ mode number, dimension, aperture diameter; depth, Secchi depth

D_n refractive index structure function

dB   decibel

dBmW   decibel milliwatts

e, q	electron charge
$E(z,r)$	spatial impulse response point spread function
E_d, E_u	downwelling, upwelling radiance at depth
E_b	energy per bit
erf	error function
erfc	complementary error function
F	F-number of lens
f	frequency (cycles/s)
$F(\mathbf{r},t)$	solution to Fresnel integral (volts/m)
$\mathbf{f}, (f_x, f_y)$	spatial frequency, bold
f_c	focal length
f_g	Greenwood frequency
f_{sg}	normalized received power
F_w	off-axis irradiance correction
FEC	forward error correction
FOV	field of view
FSL	Fraunhofer spreading loss
G	gain
g	number of indistinguishable states in ground level of the atom, asymmetry factor
h	Planck's constant height
H	scale height
$h(\mathbf{r},t)$	effect of single electron flow
$h(x,y;\omega)$	linear spatial effect at frequency ω
$H, E\ (I)$	irradiance (intensity) (watts/m^2)
HV	Hufnagel-Valley
HAP	Hufnagel-Andrews-Phillips
$I_e, I(\mathbf{R}); I(\mathbf{v})$	radiant intensity (watts/sr): image intensity 2D; image evolving in time
I_D	dark current
$I_0(..)$	modified Bessel function of the first kind with order zeo
J_0, J_1	zeroth and first-order Bessel functions
k	Boltzmann's constant (J/deg), electron count, diffuse attenuation coefficient
$\mathbf{k}, \kappa$	wavenumber vector
K	photo-electron counts
K	degrees Kelvin (degrees)
k_D, i_T	detection threshold
k_T	integer value of the detection threshold that minimizes the probability of error
ℓ_0	inner scale of turbulence
L_e	radiance (w/m^{2}sr), propagation path length
$L(z)$	normalized range equation (log of range corrected range equation $\times r^2$)
L_0	outer scale of turbulence
L_a, L_c	atmospheric, cloud transmittance
$L_{a/s}$	air/sea interface transmittance
m	mass of an atom

M_e, $M(\mathbf{r})$	radiant exitance (radiant emittance) (watts/m^2), point source coherence function, number of slots, number of symbols
M_{FM}	FM modulation index
M_L	number of local field modes
M_R	number of received modes
m_v, $m(t)$, $m(\mathbf{r})$	Poisson intensity variable, number of electron counts; time modulation, spatial modulation
m_0	total noise counts
m_1	total signal plus noise counts
MTF	modulation transfer function
n	number of atoms; index of refraction of air
n_s	average number of photo-electrons per bit for a BER $= 10^{-9}$
n_q	average number of photo-electrons per bit at the quantum limit
N_e	spectral radiance (w/m^{2}srλ); number density of air molecules, number of slots, number of Zernike modes
N_0	unilateral spectral noise density
NRZ	non-return-to-zero
OAGC	optical automatic gain control
OTF	optical transfer function
P	probability density
P_E, PE	probability of bit error
P_t	probability of t-errors
$P(x,y)$, $P(z)$	pupil function; pressure
$p(\theta)$	scattering phase function
P_e, P, $P(z)$	radiant flux (radiant power) (watts); probability; pressure, power
PIB	power in the bucket
PIF	power in the fiber
POF	power out of the fiber
PRF	pulse repetition frequency
$Q(a,b)$	Marcum Q-function
Q, Q_e, Q_s	radiant energy (joules); extinction efficiency; scattering efficiency, compensation factor
R	distance, data rate, water reflectance
RZ	return to zero
r $r(R_b)$	scalar range spectral efficiency, detector radius
$\vec{r}$	radius vector
r_c, $\mathbf{r}$	radius, radius of the aperture
r_0	Fried parameter, radius of the aperture
$R(\rho)$	covariance of homogeneous process
$R(\tau)$	covariance of wide sense stationary process
R_n^m	radial polynomial
R_e	radius of the Earth
R_L	load resistor

RR	Rayleigh range
r, s,	vector distance (arrow, bold)
S	entropy (logarithm of degrees of uncertainty – information), spectral efficiency
$s, s(I,j)$	volume scattering coefficient, signal
$S(\omega)$	power spectral density
SNR, $[\frac{S}{N}]$	signal-to-noise ratio
SR	Strehl ratio
$t, T, \mathrm{T_o}$	time, temperature
T_D	lase dead time
U	input potential to Fresnel integral (volts/m)
Var[ ]	variance
$V_{\mathrm{HV\ 5/7}}$	HV 5/7 vertical profile for wind speed
VOA	variable optical attentuator
w	wind speed, error of focus
w_g	ground wind speed
w_s	beam slew rate
$W(x,y)$	aberration function
x	Mie parameter
$x(\mathbf{r},t)$	shot noise process
z, R	range
Δ	maximum offset
ΔT	slotwidth
$\Delta \lambda$	spectral passband
Δv	Laser frequency linewidth
$\Phi_m(\mathbf{f},t)$	spatial power spectral density of m(**r**,t)
Φ_n	refractive index probability density function
$\Gamma(\mathbf{r},\mathbf{r'};t,t');$ $M(\rho,z)$	mutual coherence function
Λ	likelihood ratio
$\Theta(\ )$	optical transfer function (OTF)
Ω	solid angle
α	detector sensitivity parameter
δ	Stephan-Boltzmann constant (J-m^2 deg^{-4} s^{-1})
ε	emissivity
ε	focus error
$\varepsilon, \varepsilon_0$	permittivity, (free space)
φ	scalar potential, angular coordinate; beamwidth
$\varphi(\mathbf{\rho}), \varphi(\tau)$	eigenfunctions
Φ_c	laser signal phase angle
Φ_o	laser oscillator phase angle
φ	optical phase
η	quantum efficiency
γ_{Tx}, γ_{Rx}	transmitter, receiver transmittance

γ_{fiber}	fiber transmittance
γ_0	signal-to-noise ratio per symbol per unit bandwidth
$\Gamma(n, x)$	incomplete gamma function
λ	wavelength
$\lambda_{\tau,l}, \lambda_{s,j}, \lambda(q,l)$	eigenvalues
μ, μ_0	permeability, (free space)
$\mu_{S,B}$	average number of photo-electrons per bit from laser signal
$\mu_{N,B}$	average number of photo- electrons per bit from background radiation and dark current
$\pi_k(\theta), \xi_k(\theta)$	spherical harmonics (can be written as Legendre polynomials)
$\theta, \alpha, \varphi_s$	angular coordinate
θ_0	isoplanatic angle
ρ	reflection coefficient; bits of resolution
ρ_0	plane wave lateral coherence length
σ	cross section, standard deviation
σ_{fit}^2	residual fitting error
$\sigma_\chi{}^2$	Rytov number
$\sigma_T{}^2$	angle of arrival tilt variance (jitter)
$\sigma_\theta{}^2$	tilt variance
$\sigma_{i_{TH}}{}^2$	thermal noise variance
τ	optical thickness, time between symbol errors
τ_d	diffusion thickness
τ_0	Greenwood time period
τ_B	bit period
ω	frequency (rad/s); single-scatter albedo, radian frequency
ψ, ψ_{BB}	Wien's law, black-body equation

1   Genesis of electro-optic systems

When we decided to write this book about the design of electro-optic systems, we agreed to make it as fundamental as possible. To do this in detail would most probably make the book unwieldy. Rather, we will try to motivate all aspects of the design from fundamental principles, stating the important results, and leaving the derivations to references. We will take as our starting point the first two Laws of Thermodynamics [1]. The Three Laws of Thermodynamics are the basic foundation of our understanding of how the Universe works. Everything, no matter how large or small, is subject to the Three Laws. The Laws of Thermodynamics dictate the specifics for the movement of heat and work, both natural and man-made. The First Law of Thermodynamics is a statement of the conservation of energy – the Second Law is a statement about the nature of that conservation – and the Third Law is a statement about reaching Absolute Zero (0° K). These laws and Maxwell's equations were developed in the nineteenth century, and are the foundation upon which twentieth-century physics was founded.

Since the Laws are so obviously important, what are they?

First Law: energy can neither be created nor destroyed. It can only change forms. In any process, the total energy of the universe remains the same.

The First Law states that energy cannot be created or destroyed; rather, the amount of energy lost in a steady-state process cannot be greater than the amount of energy gained. This was a rebuff to early attempts at perpetual motion machines. This is the statement of conservation of energy for a thermodynamic system. It refers to the two ways that a closed system transfers energy to and from its surroundings – by the process of heating (or cooling) and the process of mechanical work. Whether the system is open or closed, all energy transfers must be in balance.

Second Law: the entropy of an isolated system not in equilibrium will tend to increase over time, approaching a maximum value at equilibrium.

The second law states that energy systems will always increase their entropy rather than decrease it. In other words heat can spontaneously flow from a higher-temperature region to a lower-temperature region, but not the reverse. (Heat can be made to flow from cold to hot, as in a refrigerator, but requires an external source of power, electricity.)

A way of looking at the second law for non-scientists is to look at entropy as a measure of disorder. A broken cup has less order than before being broken. Similarly solid crystals have very low entropy values, while gases which are highly disorganized have high entropy values.

1.1   Energy

What does this have to do with optics? Well, about the same time that the thermodynamic laws were being developed, the laws of electricity were also being developed. James Clerk Maxwell unified the Laws of Gauss, Ampere and Faraday into the set of laws now referred to as Maxwell's equations [2]. These equations show that electric and magnetic fields are coupled and satisfy a wave equation,

$$\nabla^2 \phi = \frac{1}{c^2} \frac{\partial^2 \phi}{\partial t^2} \tag{1.1}$$

where ϕ can represent the scalar components of the electric and magnetic fields. In a vacuum the speed of light $c = 1/\sqrt{\varepsilon_0 \mu_0}$ where ε_0, μ_0 are the permittivity and permeability of free space, as derived in Maxwell's equations. Since electromagnetic fields can propagate in a vacuum, no loss occurs (unless obstacles are encountered) and the fields can propagate indefinitely since energy can be neither created nor destroyed. Thus the energy from the Sun can bring life to the planet Earth, and we can see the vast vista of stars at night, even though they are light years away. A general solution to the wave equation (in spherical coordinates) for the fields takes the form

$$\phi = \text{Constant} \times \frac{f\left(t - \frac{r}{c}\right)}{r}. \tag{1.2}$$

$t - \frac{r}{c}$ is referred to as the retarded time. That is, the waveform of the field propagates undistorted, but attenuated (spread out in space) by the factor r, when traveling the distance r, in a time r/c. Since the energy in the fields is proportional to $|\phi|^2$ [2], we see that the energy in a propagating electromagnetic field falls off as $1/|r|^2$. From this we can deduce several properties regarding the transfer of energy by electromagnetic means. For the purpose of this book we will use the following standard notation for Radiant quantities:

> **Radiant energy** is **energy** carried from any electromagnetic field. It is denoted by Q_e. Its **SI** unit is the **joule** (J).
>
> **Radiant flux** is radiant energy per unit time (also called **radiant power**); it is considered the fundamental radiometric unit. It is denoted by P_e. Its **SI** unit is the **watt** (W).
>
> **Radiant exitance**, or **radiant emittance**, is radiant flux emitted from an extended source per unit **source area**. It is denoted by M_e. Its **SI** unit is: watt per square meter (W/m^2).
>
> **Irradiance** is radiant flux incident on a surface unit area. It is denoted by E_e. Its **SI** unit is: **watt per unit area** (W/m^2).
>
> **Radiant intensity** is radiant flux emitted from a point source per unit **solid angle**. It is denoted by I_e. Its **SI** unit is: watt per **steradian** (W/sr).
>
> **Radiance** is radiant flux emitted from an extended source per unit **solid angle** and per unit **projected source area**. It is denoted by L_e. Its **SI** unit is: watt per steradian and square meter (W/(sr m^2)).
>
> **Spectral** in front of any of these quantities implies the same **SI** unit per unit **wavelength**.

Thus spectral radiant energy is the energy radiated per unit wavelength. These noise sources generally are not described analytically, and depending upon the instrument used fall into one of the categories described above.

1.2 The range equation

We will try to show how these terms are used, with a few examples.

Example 1.1 Suppose a point source radiated (transmitted) an amount of radiant power, $P_t = Q_e$ joules/second, equally into all directions (4π steradians or omnidirectional). Then the radiant intensity would be equal to

$$I_t = \frac{P_t}{4\pi} \tag{1.3}$$

A receiver with an area A_r located a distance r from the point source would subtend the solid angle $A_r/|r|^2$ to the source. Therefore such a receiver could capture an amount of radiant power (power) equal to Figure 1.1,

$$P_r = \frac{P_t A_r}{4\pi |r|^2} \tag{1.4}$$

which conforms to the $1/|r|^2$ requirement of a propagating field. Suppose that a second identical source could direct all its energy toward a receiver, uniformly into a solid angle $\Omega < 4\pi$. Such a source would then transmit $4\pi/\Omega$ times more power to the receiver, without the receiver knowing that it wasn't a source with $(4\pi/\Omega)P_t$ watts. For this reason $4\pi/\Omega$ is called the gain (G_t – over isotropic) of the second system, and we write the received power as

$$P_r = \frac{P_t A_r G_t}{4\pi |r|^2} \tag{1.5}$$

This is referred to as the frequency-independent form of the range equation, and defines the transfer of power in a radiating system.

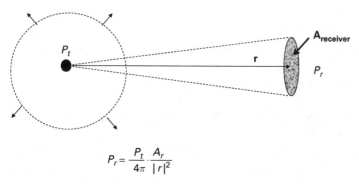

$$P_r = \frac{P_t}{4\pi} \cdot \frac{A_r}{|r|^2}$$

Figure 1.1 Range equation in geometric form.

Example 1.2 How can we calculate the total power emitted by the Sun? The Sun is located a distance $R_{sun} = 92 \times 10^6$ miles from the Earth. The irradiance of the Sun at the top of the Earth's atmosphere (solar constant), I_{sun}, is approximately 1500 w/m^2. The Earth has a cross section, A_{earth}, of $\pi(4000)^2 = 50.3 \times 10^6$ square miles, or a cross section of 1.3×10^{14} m^2. Thus the Earth intercepts $I_{sun}A_{earth}$ watts from the Sun, which is approximately 1.96×10^{17} watts. The solid angle subtended from the Earth to the Sun is $\Omega_{sun-earth} = A_{earth}/R_{sun}^2$ steradians, so the Earth intercepts the percentage $\Omega_{sun-earth}/4\pi$ of the Sun's emission, or only 5.81×10^{-7} per cent of the energy radiated by the sun. This means that the sun emits a total of $\left(4\pi/\Omega_{sun-earth} \right)I_{sun}A_{earth} = 3.37 \times 10^{25}$ watts.

1.3   Characterization of noise

First we will relate all the radiant quantities described earlier. We start first with the most elemental of the quantities, the spectral radiance, $N_e(r,\theta,\lambda)$. To obtain the radiance, we would perform the integration, which for narrowband filters can be approximated, as follows:

$$L_e(r,\Omega) = \int_{\lambda_1}^{\lambda_2} N_e(r,\Omega,\lambda)d\lambda \approx N_e(r,\Omega,\lambda)\Delta\lambda \tag{1.6}$$

From $L_e(r,\Omega)$ we can go in one of two directions. To obtain the radiant emittance we perform the integral

$$M_e(r) = \int_0^{\Omega_0} L_e(r,\Omega)d\Omega \approx L_e(r,\Omega)\Omega_0 = N_e(r,\Omega,\lambda)\Omega_0\Delta\lambda \tag{1.7}$$

while to get the radiant intensity, we perform the integral

$$I_e(\Omega) = \int_A L_e(r,\Omega)dS \approx L_e(r,\Omega)A = N_e(r,\Omega,\lambda)A\Delta\lambda \tag{1.8}$$

Finally, to obtain the power, we perform either one of two integrals:

$$P_e = \int_A M_e(r)dS \approx M_e(r)A = L_e(r,\Omega)A\Omega_0 = N_e(r,\Omega,\lambda)A\Omega_0\Delta\lambda \tag{1.9}$$

or

$$P_e = \int_0^{\Omega_0} I_e(\Omega)d\Omega \approx I_e(\Omega)\Omega_0 = L_e(r,\Omega)A\Omega_0 = N_e(r,\Omega,\lambda)A\Omega_0\Delta\lambda \tag{1.10}$$

We have defined I_e to be from a point source, but by allowing a variation with angle it becomes more general. A point source would have no variation with angle. We have also used Ω to denote (θ, ϕ) and, $d\Omega$ the Jacobian. The Sun, which has a diameter of 863 000 miles, only subtends 0.5 degrees to Earth, and for many cases appears like a point source, but can project different values in different directions. We have used the description of frequency and bandwidth in customary terms of wavelength and differential wavelength, because these sources were first studied by astronomers, and that is how they were described. We also have the relationship between frequency and wavelength of $f = c/\lambda$ with $df = |(c/\lambda^2)|d\lambda$.

Example 1.3 Let us assume that a source radiates a spectral radiance $N_\lambda (w/m^2 - sr - BW)$. Suppose that the receiver has a field of view (FOV – the solid angle that it views) equal to Ω_{rec}, a collection area A_{rec}, and a filter bandwidth $\Delta\lambda$. If R is the range from the source to the receiver, then integrating over the source area yields (1) $\Omega_{rec}R^2$ if the source fills the FOV, or (2) A_{source} if it doesn't. The receiver subtends the solid angle A_{rec}/R^2 to the source. Therefore the total power collected by the receiver becomes

$$N_\lambda \Omega_{rec} R^2 \frac{A_{rec}}{R^2} \Delta\lambda = N_\lambda \Omega_{rec} A_{rec} \Delta\lambda \qquad (1.11)$$

if the source fills the receiver FOV. This would be in cases like imaging systems, where the FOV of each detector pixel fills the cell in the image that is being viewed. On the other hand in a detection system, such as a system searching for a hot target, the FOV might be larger than the target and

$$N_\lambda \frac{A_{source}}{R^2} A_{rec} \Delta\lambda \qquad (1.12)$$

would most likely apply.

1.4   Black-body radiation

Black-body or "pure-temperature" radiation was the name given to radiation emanating from a system in thermal equilibrium. This is the fundamental source of primary (suns, stars, plasmas, etc.) and secondary (moons, planets, atmosphere, etc.) electromagnetic noise, and was the exciting area of research in thermodynamics. In addition, the outcome proved fundamental to the development of the quantum theory.

It was known that in a sealed chamber at a given temperature, the spectral distribution of radiation is independent of the material out of which the chamber is made. In 1884, Boltzmann [3] showed that the total irradiance energy in this distribution varied as the fourth power of the temperature. The constant of proportionality, $\delta = 0.567 \times 10^{-8}$ J–m^{-2} deg^{-4} s^{-1}, is known as the Stefan-Boltzmann constant,

$$\text{Irradiance} \sim \delta T_0^4 \qquad (1.13)$$

and the law as the Stefan-Boltzmann law. In 1893 Wien [4] showed that this functional dependence must be of the form

$$\psi_\lambda \sim T_0^5 f(\lambda T_0) = \frac{1}{\lambda^5} F(\lambda T_0) \qquad (1.14)$$

where T_0 is the temperature in degrees Kelvin. He assumed a form

$$\psi_\lambda \sim \lambda^{-5} e^{k/\lambda T_0} \tag{1.15}$$

called Wien's law. He also developed Wien's displacement law, which states that if the product λT_0 is held constant, ψ_λ / T_0^5 is the same at all temperatures.

In 1900, Rayleigh [5] made a suggestion which was based on a technique used successfully in statistical mechanics. This was followed up with a calculation made by Jeans, resulting in the Rayleigh-Jeans formula. The idea relied on the equipartition of energy, and went as follows: one can expand the electromagnetic field contained in an enclosure into orthogonal and independent modes (degrees of freedom). By then associating the total energy kT_0 with each degree of freedom (used in the ideal gas law) an expression for the radiation can be obtained. Jeans calculated the number of modes per cycle to be D_{BB}, where

$$D_{BB} = \frac{8\pi f^2}{c^3} \tag{1.16}$$

yielding the expression for the energy to be

$$E = kT_0 D_{BB} = \frac{8\pi f^2 kT_0}{c^3}. \tag{1.17}$$

Unfortunately, this predicted an infinite energy instead of the fourth-power dependence on T_0. The two expressions E and ψ_λ, are plotted in Figure 1.2, along with precise

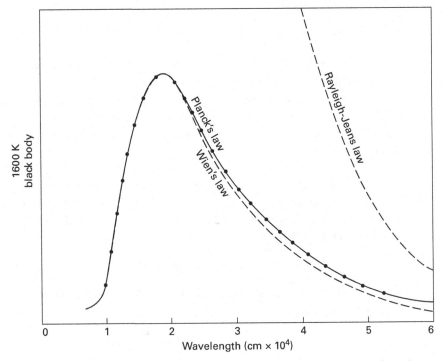

Figure 1.2 Black-body radiation, Wien's law, and the Rayleigh-Jeans approximation. The experimental curve is fit exactly by Planck's law.

measurements taken by Lummer and Pringsheim [6]. Neither was correct, but Wien's law was asymptotically correct at high frequencies, and the Rayleigh-Jeans law was asymptotically correct at low frequencies.

This was the picture as seen by Planck around 1900. Plank's derivation of the correct equation for the black-body law came through his understanding of thermodynamics. He curve-fitted the second derivative of entropy with respect to energy to yield the correct two asymptotic forms, and worked them back to obtain the energy equation [7]. The most commonly used derivation (suggested by Einstein) multiplies the mode density D_{BB} by the quantum mechanical equivalent of kT_0, which is

$$\frac{hf}{e^{hf/kT_0} - 1} \tag{1.18}$$

Thus we see that the equation for black-body radiation can be derived to be

$$\psi_\lambda = \frac{8\pi f^2}{c^3} \frac{hf}{e^{hf/kT_0} - 1}$$

$$= \frac{8\pi hc}{(\lambda)^5} \frac{1}{e^{hc/k\lambda T_0} - 1} \tag{1.19}$$

which fits the energy density curves exactly.

This is the energy per unit of wavelength (bandwidth) contained in a sealed container. When we consider the radiation from an isotropic black body, we must multiply the energy by the speed of light, c, divided by 4π, for isotropy, and divide by two for each polarization. This yields

$$\psi_\lambda \frac{c}{4\pi} \frac{1}{2} = \frac{8\pi f^2}{c^3} \frac{hf}{e^{hf/kT_0} - 1} \frac{c}{4\pi} \frac{1}{2} = \frac{f^2}{c^2} \frac{hf}{e^{hf/kT_0} - 1} \tag{1.20}$$

which has the dimensions of energy per unit bandwidth per unit solid angle per unit area. As such it is a spectral radiance, N_{BB}. Reflecting back on example 1.3, the Sun appears as a 5778° black body. If the receiver field of view is greater than 0.5 degrees we can treat example 1.2 as before (example 1.3b). If the field of view is less than 0.5 degrees we are only looking at a small cell in the Sun, which in turn fills the field of view of the pixel viewing it (example 1.3a). When an instrument measures the solar irradiance, E_{sun}, its field of view is larger than that subtended by the Sun and hence measures

$$E_{sun} = N_\lambda \frac{A_{sun}}{R^2}, \tag{1.21}$$

which has dimensions of joules/μm-m² as given in example 1.2.

The quantum mechanical equipartition energy has the two asymptotic forms. At the low energy (low frequency), $hf < kT_0$, we have

$$\frac{hf}{e^{hf/kT_0} - 1} \approx \frac{hf}{1 + \frac{hf}{kT_0} - 1} = kT_0 \qquad (1.22)$$

the classic value. At the high end we have $hf \exp(-hf/kT_0)$, which approaches Wien's law.

What Planck also observed about black-body radiation was that this described a sum of harmonic oscillators,

$$\frac{8\pi f^2}{c^3} \frac{hf}{e^{hf/kT_0} - 1} = \frac{8\pi f^2}{c^3} \frac{hf}{e^{hf/kT_0}\left(1 - e^{-hf/kT_0}\right)} = \frac{8\pi hf^3}{c^3 e^{hf/kT_0}} \sum_{n=0}^{\infty} e^{-nhf/kT_0} \qquad (1.23)$$

which was the first quantum hypothesis.

We have expanded on the derivation of black-body radiation because it is so important in understanding the composition of background noise. This will become clearer as we see the relevance of modes, mode number and energy per mode.

We show several representative curves of astronomical radiation in Figures 1.3–1.10. In Figure 1.3, we show the approximate black-body radiation of the Sun outside the atmosphere, and also modified by the absorption bands of the atmosphere [8]. In Figure 1.4, we show this in logarithmic form as a function of the zenith angle. Notice that the units are different, but the results are the same [8]. In Figure 1.5, we show the sky radiance during the day, which

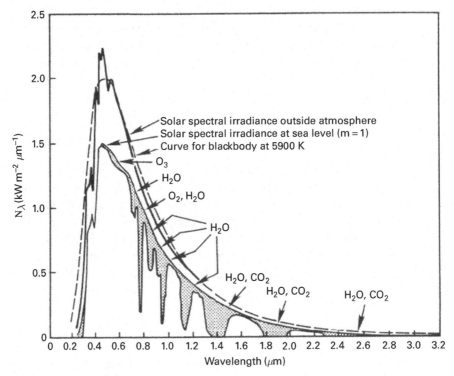

Figure 1.3 Spectral distribution curves related to the Sun. The shaded areas indicated absorpion at sea level due to the atmospheric constituents shown.

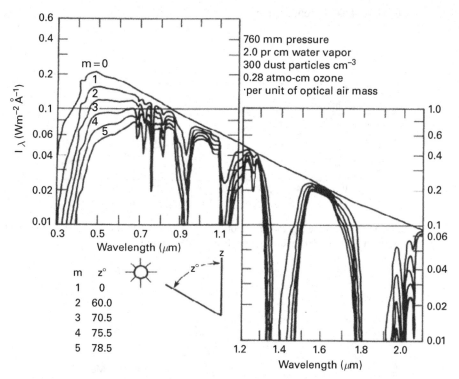

Figure 1.4 Solar spectral irradiance curves at sea level for various optical air masses. The value of the solar constant used in this calculation was 1322 W/m² [23].

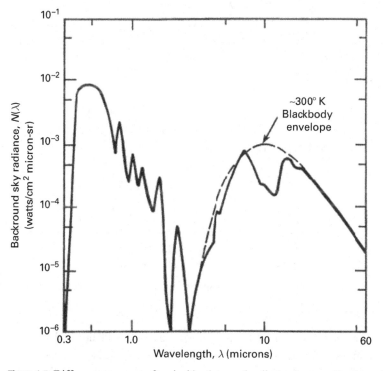

Figure 1.5 Diffuse component of typical background radiance from sea level; zenith angle, 45°; excellent visibility [23].

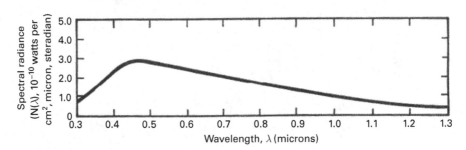

Figure 1.6 Nighttime sky radiance from zenith due to zodiacal light, galactic light, and scattered starlight [23]. (Courtesy of National Bureau of Standards.)

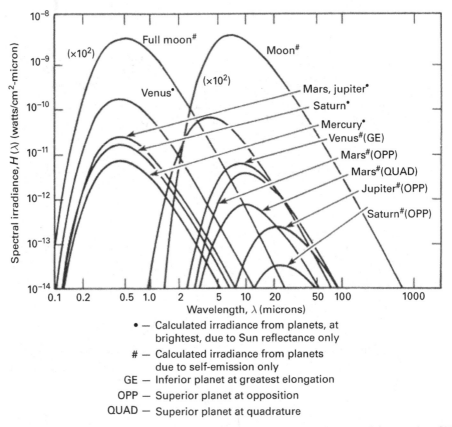

Figure 1.7 Calculated planetary and lunar spectral irradiance outside the terrestrial atmosphere [23].

consists of the re-radiation of the solar energy plus the re-radiation of the Earth's emission at approximately 300° Kelvin [9]. In Figure 1.6, we see the sky radiance at night [10]. In Figure 1.7, we show the irradiance from the planets, Moon and Venus [11]. In Figure 1.8, we see their spectral albedo (spectral albedo is defined as the ratio of the total reflected energy

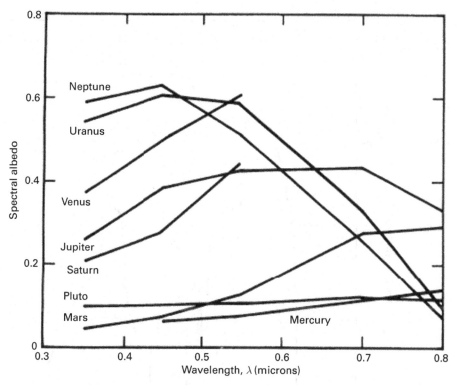

Figure 1.8 Spectral albedo as a function of wavelength [23].

from a surface to the total incident energy on said surface) [12]. In Figure 1.9, we see the spectral radiant emittance of the Earth as viewed from space [13]. Finally in Figure 1.10 we see the spectral irradiance of the brightest stars. A further discussion of these curves can be found in [14]. For an intensive description of the development of the black-body theory and its link to the development of quantum mechanics see Whittaker [15].

1.5   Entropy

As the Second Law shows, systems tend towards a balance of disorder, with entropy always increasing. Thus if all air conditioning and heating were turned off in a house, the temperature would eventually equal that of its surroundings. While it is possible to build a machine that can raise (or lower) the house temperature, these mechanical systems can at best make a perfect exchange of heat. In fact, all mechanical systems have some inefficiency; hence some waste energy is generated. (And in fact, the universe is constantly losing usable energy.)

Entropy is defined as the measure of uncertainty in a system. For the case of an ideal monatomic gas in a container of volume V that is thermally insulated from its surroundings, the entropy can be expressed using the Sackur-Tetrode equation [16] by

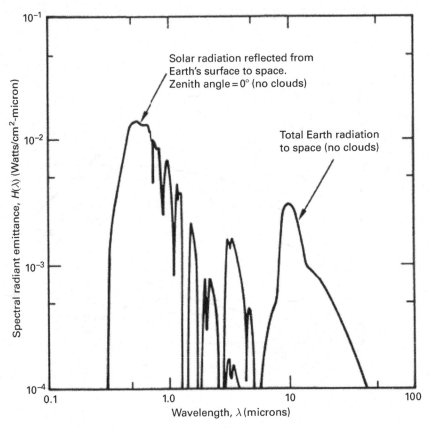

Figure 1.9 Spectral radiant emittance of the earth [23].

$$S = kn\left[\frac{5}{2} + \ln\left(g\frac{V}{n}\left\{\frac{4\pi mE}{3h^2n}\right\}^{3/2}\right)\right] \tag{1.24}$$

where

n = number of atoms
m = mass of an atom
k = Boltzmann's constant
h = Planck's constant
E = total energy
g = number of indistinguishable states in the ground level of the atom.

If the ground state is not degenerate, $g = 1$. This is the case of an atom with no momentum in the ground state ($j = 0$). If there is a moment of momentum j (spin plus orbit), there are $g_j = 2j + 1$ states, which are supposed to be equally probable. If we have spin $j = 1/2$ (two orientations), then $g = 2$. For $j = 1, g = 3$ and three orientations ($-1,0,+1$). If we compute S_1 for $g = 1$, then $S_2 = S_1 + kn(\ln 2)$ and $S_3 = S_1 + kn(\ln 3)$. The importance of this example is that we see the entropy is proportional to the logarithm of the possible orientations of the atom.

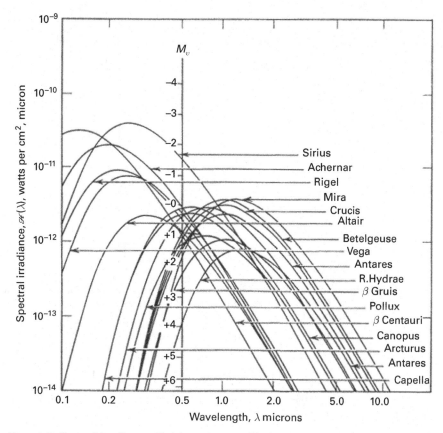

Figure 1.10 Spectral irradiance of brightest stars outside the terrestrial atmosphere [24].

Information too can be written as the logarithm of the total amount of uncertainty. This idea had been used by Nyquist [17] and Hartley [18].

Example 1.4 Suppose we had to guess a number that existed between zero and 31 ($32 = 2^5$ possibilities), and suppose we could ask n yes/no questions. How many questions would it take to find the number? The first question should be "is it between zero and 15"? If the answer is yes, we ask whether the number is between zero and seven. If the answer is no, we ask whether the number is between 8 and 11. And so forth. If, for example, the number was 27 we would get a no first, a no second, a yes third and two more no's. If we assign zero to a yes and a one to a no, we would get 11011, or the binary word for 27. Thus we see that information is the negative of entropy [19]. If we assigned one bit of information to each binary digit we would have five bits of information. We see that the total amount of information would be $\log_2(32) = 5$ bits; Figure 1.11. If we were to pick two numbers independently between zero and 31, we would now have $32 \times 32 = 1024$ possibilities. Since $\log_2(1024) = 10 = 5 + 5$, we see that information is additive as it should be.

Example: Suppose we had to guess a number that existed between zero and 31 (32 = possibilities), and suppose we could ask C yes/no questions. Assign a 1 to no and zero to yes. How many questions would it take to find the number?

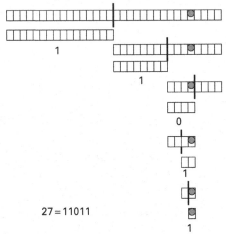

If we were to pick two numbers independently between zero and 31, we would now have 32×32 = 1024 possibilities. Since $\log_2(1024) = 10 = 5 + 5$, we see that information is additive as it should be.

Figure 1.11 Total amount of information for Example 1.4.

The choice of the base is merely a question of units, since $\log_b n = (\log_b c)\log_c n$. Thus

$$\log_{10} n = (\log_{10} 2)\log_2 n = (0.301)\log_2 n \text{ in Hartleys,}$$

and

$$\log_e n = (\log_e 2)\log_2 n = (0.693)\log_2 n \text{ , in nats (natural units).}$$

However, Shannon applied this concept to a signal space defined by a dimension $2BT$ (defined as the number of Nyquist samples of a function band limited to B, and observed over the time T), a signal power P, and a noise power N [20]. Then using the definition of capacity, C, as

$$C = \frac{\log_2 M}{T} \tag{1.25}$$

where M is the possible number of distinct band limited signals that could exist in this space in the time T, he derived the Capacity Theorem as

$$C = B\log_2 \frac{P+N}{N} = B\log_2\left(1 + \frac{P}{N}\right) \text{ bits per second} \tag{1.26}$$

In his second ground-breaking theorem, he showed that codes existed which could achieve this capacity with arbitrarily small error.

What is important to this discussion is how he defined P and N. These were both defined as the variance (signal power), P, of the signal and variance N of additive Gaussian noise. (He also showed that Gaussian noise was the worst-case noise, and hence this was a lower bound on capacity for any additive noise.)

From Fourier analysis we know that the variance, co-variance and power spectral density are inter-related. To clarify this we first see that the mean of a function $f(t)$ is

$$E[f(t)] = E[f(t+\tau)] = \bar{f} \tag{1.27}$$

[21]. The covariance of the function is defined as

$$R_f(\tau) = E[(f(t+\tau) - \bar{f})(f(t) - \bar{f})^*]$$
$$= E[f(t+\tau)f^*(t)] - \bar{f}^2 \tag{1.28}$$

Since $\bar{f}$ is a constant, in engineering terms it represents the DC part of the signal and has no bandwidth. However, $E[f(t+0)f^*(t)] = E[|f(t)|^2] = R_f(0)$. Consequently, since the Fourier transform of $R_f(\tau)$ is the power spectral density, $S_f(\omega)$,

$$S_f(\omega) = \int\limits_{-\infty}^{\infty} R_f(\tau)e^{-j\omega\tau}d\tau \tag{1.29}$$

we see that this represents the AC portion of the spectrum. Furthermore

$$R_f(0) = \frac{1}{2\pi} \int\limits_{-\infty}^{\infty} S_f(\omega)d\omega = E[f(t)f^*(t)] = P \tag{1.30}$$

is the variance of the signal as defined in Shannon's theorem. Similarly for the noise, $n(t)$ with covariance $R_n(\tau)$ and power spectral density $S_n(\omega)$

$$R_n(0) = \frac{1}{2\pi} \int\limits_{-\infty}^{\infty} S_n(\omega)d\omega = E[n(t)n^*(t)] = N \tag{1.31}$$

Consequently, the correct interpretation of the signal-to-noise ratio in the capacity formula is

$$\text{SNR} = \frac{\dfrac{1}{2\pi} \int\limits_{\substack{\text{signal} \\ \text{bandwidth}}} S_f(\omega)d\omega}{\dfrac{1}{2\pi} \int\limits_{\substack{\text{signal} \\ \text{bandwidth}}} S_n(\omega)d\omega} = \frac{R_f(0)}{R_n(0)} \tag{1.32}$$

When considering electro-optical communication systems this definition applies directly.

There is also an important linkage to be made between energy and information. Suppose we express the power P in terms of energy and time as $P = E/T$. We will also consider a one-sided spectral white noise spectral density of N_0. We can then recast the capacity equation as

$$C = \frac{P}{N_0} \left[\frac{N_0 B}{P} \log_2 \left(1 + \frac{P}{N_0 B} \right) \right] \tag{1.33}$$

or

$$\frac{C}{B} = \frac{P}{N_0} \left[\frac{N_0}{P} \log_2 \left(1 + \frac{P}{N_0 B} \right) \right] \tag{1.34}$$

In the limit as the bandwidth goes to infinity, we have

$$\lim_{B \to \infty} \frac{C}{B} = \lim_{B \to \infty} \frac{P}{N_0} \left[\frac{N_0}{P} \log_2 \left(1 + \frac{P}{N_0 B} \right) \right] \tag{1.35}$$

Using the identity

$$\lim_{x \to \infty} x \log_2 \left(1 + \frac{1}{x} \right) = \log_2 e \tag{1.36}$$

[22], we can rewrite the above limit of the capacity as

$$\lim_{B \to \infty} \frac{C}{B} = \frac{P}{N \ln 2} \tag{1.37}$$

where the $N = N_0 B$ equals the noise power. This equation implies that for a fixed SNR, the normalized channel capacity approaches a maximum in the limit of infinite bandwidth. It also implies that for the channel capacity to increase, either the transmit power must increase, or the unilateral noise power density must decrease. Finally, by normalizing the received energy in terms of energy per bit,

$$E_b = \frac{P}{C}, \tag{1.38}$$

we can derive the minimum value of E_b/N_0 as ln 2.

Alternatively, we can then recast the capacity equation as

$$\frac{C}{B} = \log_2 \left(1 + \frac{E}{N_0 BT} \left(\frac{C}{C} \right) \right) = \log_2 \left(1 + \frac{1}{N_0} \left(\frac{E}{CT} \right) \left(\frac{C}{B} \right) \right). \tag{1.39}$$

Here we express the power P in terms of energy and time as $P = E/T$. Dimensionally, E/CT is energy per bit, E_b, and B/C is cycles per bit. This equation can then be recast as

$$\frac{e^{(C/B)\ln 2} - 1}{C/B} = \frac{E_b}{N_0} \tag{1.40}$$

which is shown in Figure 1.12. In the limit as $\frac{C}{B}$ goes to zero, i.e., the number of cycles per bit goes to infinity, the energy per bit approaches ln 2. For thermal noise $N_0 = kT_0$, so we see that the minimum amount of energy needed to transmit a bit of information

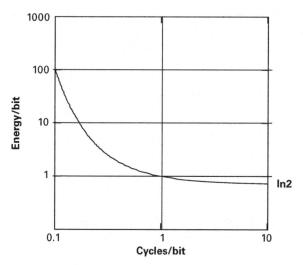

Figure 1.12 Energy efficiency in joules per bit versus cycles per bit.

is $kT_0 \ln 2$ joules. However, as we try to increase the number of bits per cycle we pay a penalty in energy per bit.

1.6 Summary

In this chapter we have tried to show the reader how the laws of thermodynamics can be used to establish the basic rules of the radiative transport of energy, and of their relationship to information theory. In particular, we have shown how these laws can be used to develop rules regarding the meaning and bounds on how one can communicate in this environment. In the next chapter we will focus more closely on the rules governing electromagnetic propagation.

References

1. M. Pidwirny. Laws of Thermodynamics. In *Fundamentals of Physical Geography*, 2nd edn (2006).
2. M. Born and E. Wolf. *Principles of Optics: Electromagnetic Theory of Propagation, Interference and Diffraction of Light*, 7th edn. Cambridge University Press, Cambridge (1999).
3. L. Boltzmann. *Ann. d. Phys.*, **xxii** (1884) pp. 31, 291.
4. W. Wien. *Ann. d. Phys.*, **lviii** (1896), p. 662.
5. Lord Rayleigh. *Phil. Mag.*, **xlix** (1900), p. 539.
6. O. Lummer and E. Pringsheim. *Verh. deutsch. Phys. Ges.*, **ii** (1900), p 163.
7. M. Planck. *Verh. deutsch. Phys. Ges.*, **Ii** (1900), p. 237.
8. W. J. Wolfe and G. J. Zissis, eds. *The Infrared Handbook*. The Environmental Institute of Michigan, Ann Arbor, MI (1978).

9. Parametric Analysis of Microwave and Laser Systems for Communication and Tracking. Hughes Aircraft Co., Report No. P67-09 (December 1966).

10. F. E. Roach. *Manual of Photometric Observations of the Airglow During the IGY.* National Bureau of Standards Report.

11. R. C. Ramsey. Spectral irradiance from stars and planets above the atmosphere from 0.1 to 100 microns. *Applied Optics*, **1**(4) (July 1962), pp. 465–471.

12. G. E. Kuiper and B. M. Middlehurst, ed. *The Solar System, Vol. III, Planets and Satellites.* University of Chicago Press (1961).

13. I. L. Goldberg. Radiation From Planet Earth. U.S. Army Signal Research and Development Laboratory Report 2231, AD-266-790 (September 1961).

14. R. M. Gagliardi and S. Karp. *Optical Communications.* Wiley Interscience, New York (1976).

15. E. Whittaker. *A History of the Theories of Aether and Electricity*, Vols. I & II. Harper and Brothers, New York (1951).

16. F. T. S. Yu. *Entropy and Information Optics.* Marcel Dekker (2000) p. 103.

17. H. Nyquist. Certain factors affecting telegraph speed. *Bell System Technical Journal*, **3** (April 1924), p. 324.

18. R. V. L. Hartley. The transmission of information. *Bell System Technical Journal*, **3** (July 1928), pp. 535–564.

19. T. Carter. An Introduction to Information Theory and Entropy. Complex Systems Summer School, Santa Fe (June 2007), pp. 15–20.

20. C. E. Shannon. A mathematical theory of communication. *Bell System Technical Journal*, **27** (July and October 1948), pp. 379–423 and 623–656.

21. A. Papoulis. *Probability, Random Variables, and Stochastic Processes.* McGraw-Hill Book Co. (1965) pp. 300–304.

22. B. P. Lathi. *Random Signals and Communications Theory.* International Textbook Company (1968) p. 471.

23. S. Karp, R. M. Gagliardi, S. E. Moran, L. B. Stotts. *Optical Channels.* Plenum Press, New York (1988).

24. W. K. Pratt. *Laser Communication Systems.* John Wiley & Sons, New York (1969).

2 Role of electromagnetic theory in electro-optics systems

In this chapter we will present the Kirkhoff-Fresnel solution to the Maxwell wave equations, (1.1). A general discussion of the solution to Maxwell's equation can be found in [1–4]. We will not attempt to present these derivations here. Rather we will present the relevant equations and discuss them in the context of electro-optic system design. In particular we will try to show the correspondence between the radiant quantities described in Chapter 1 and the second-order moments of the electromagnetic field. These are the irradiance and radiant flux as they relate to intensity and power in the field.

2.1 Kirchhoff diffraction theory

If we assume a monochromatic signal of the form

$$\phi = f(\mathbf{r}; \omega)e^{-j\omega t} \tag{2.1}$$

then the wave equation, Eq. (1.1), reduces to the Helmholtz equation given by

$$\left(\nabla^2 + k^2\right) f(\mathbf{r}; \omega) = 0 \tag{2.2}$$

where $k = \omega/c$. If we take the general solution $F(r,t)$ of the wave equation, we have

$$F(\mathbf{r}, t) = \int_{-\infty}^{\infty} f(\mathbf{r}, \omega)e^{-j\omega t} \frac{d\omega}{2\pi} \tag{2.3}$$

with

$$f(\mathbf{r}, \omega) = \int_{-\infty}^{\infty} F(\mathbf{r}, t)e^{j\omega t} dt \tag{2.4}$$

Using the Helmholtz equation and Green's theorem [1–3] one can derive the Fresnel-Kirchhoff diffraction formula for the electromagnetic field incident on a receiver as

$$U(P_r) = -\frac{Cj\omega}{2c} \int_{Surface} \frac{e^{jk(r+s)}}{rs} [\cos(\mathbf{n}, \mathbf{r}) - \cos(\mathbf{n}, \mathbf{s}) dS] \tag{2.5}$$

where the geometry is described in Figure 2.1 with the source at P_0, and the receiver at P. The integration is taken over the closed surface A+B+C, with the only finite contribution to the integral being over the surface A, the receiving aperture. The dimensions of this field are volts per unit length per radian.

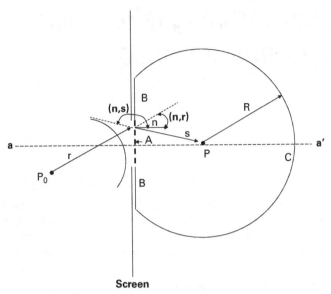

Screen

Figure 2.1 Geometry for Kirchhoff diffraction theory.

This is the geometry for an arbitrary spherical wave emanating from P_0 and observed at point P. As we make the distance r grow large and place the points P, P_0 on the a–a' axis that is normal to the screen, the illumination over the aperture A becomes planar. In addition the directional cosines (n,r), (n,s), approach 1 and -1 respectively. This then reduces to the integral

$$U(P_r) = -\frac{j\omega}{c} \int_{Surface} f(\mathbf{r}, \omega)\frac{e^{j\mathbf{k}\cdot\mathbf{r}}}{r} dS \tag{2.6}$$

where $f(\mathbf{r}, \omega)$ is the amplitude over the aperture at the frequency ω. The general solution would then be

$$F(\mathbf{r},t) = \int_{-\infty}^{\infty} U(P_r)e^{-j\omega t}\frac{d\omega}{2\pi} = \int_{-\infty}^{\infty} -\frac{j\omega}{c}e^{-j\omega t}\frac{d\omega}{2\pi}\int_{Surface} f(\mathbf{r}, \omega)\frac{e^{j\mathbf{k}\cdot\mathbf{r}}}{r} dS$$

$$= \int_{Surface}\frac{dS}{r}\int_{-\infty}^{\infty} -j\omega f(\mathbf{r}, \omega)e^{-j(\omega t-\mathbf{k}\cdot\mathbf{r})}\frac{d\omega}{2\pi} = \int_{Surface}\frac{dS}{r}\frac{d}{dt}\left[f\left(t-\frac{r}{c}\right)e^{-j(\omega t-\mathbf{k}\cdot\mathbf{r})}\right]$$

$$\tag{2.7}$$

The dimensions of $F(\mathbf{r},t)$ are volts per unit length. For narrowband solutions, engineers use the approximation [4]

$$F(r,t) = \int_{-\infty}^{\infty} U(P_r)e^{-j\omega t}\frac{d\omega}{2\pi} = \int_{-\infty}^{\infty} -\frac{j\omega}{c}e^{-j\omega t}\frac{d\omega}{2\pi}\int_{Surface} f(\mathbf{r}, \omega)\frac{e^{j\mathbf{k}\cdot\mathbf{r}}}{r} dS$$

$$\approx \int_{Surface}\frac{dS}{r}(-j\tilde{\omega})\int_{-\infty}^{\infty} f(\mathbf{r}, \omega)e^{-j(\omega t-\mathbf{k}\cdot\mathbf{r})}\frac{d\omega}{2\pi}$$

$$= (-j\tilde{\omega})\int_{Surface}\frac{dS}{r}f\left(t-\frac{r}{c}\right)e^{-j(\omega t-\mathbf{k}\cdot\mathbf{r})} \tag{2.8}$$

In all these cases we have used r as the scalar distance from points on the aperture to points at the receiver. The general solution can be seen to be an integration of differential "Huygens wavelets". The narrowband solution is actually used by communications engineers when modeling multi-path. This model is generally a good approximation on axis. However, when looking off-axis with wide bandwidths, distortions can occur in addition to the deleterious effects of propagation, and the full solution should be used.

If we assume Cartesian coordinates, and denote the aperture plane by (x_1, y_1) and the receiver plane by (x_0, y_0), the scalar distance r between any point in the aperture plane and any distance in the receiver plane becomes

$$r = \sqrt{z^2 + (x_0 - x_1)^2 + (y_0 - y_1)^2}$$

$$= z \left(\sqrt{1 + \left(\frac{x_0 - x_1}{z}\right)^2 + \left(\frac{y_0 - y_1}{z}\right)^2} \right)$$

$$\cong z \left[1 + \frac{1}{2}\left(\frac{x_0 - x_1}{z}\right)^2 + \frac{1}{2}\left(\frac{y_0 - y_1}{z}\right)^2 \right] \tag{2.9}$$

for large z. The solution to the Fresnel-Kirkoff diffraction formula comes in two basic regions. The close-in region is called the Fresnel region.

2.2   The Fresnel approximation

In the Fresnel region, defined by

$$z^3 >> \frac{\pi}{4\lambda} \left[(x_0 - x_1)^2 + (y_0 - y_1)^2 \right]^2_{max} \tag{2.10}$$

the approximation becomes

$$U(x_0, y_0 : z) = \frac{1}{j\lambda z} \int_{Surface} f(x_1, y_1 : \omega) e^{jk\left[(x_0-x_1)^2 + (y_0-y_1)^2 + z^2\right]^{1/2}} dx_1 dy_1$$

$$\cong \frac{e^{jkz}}{j\lambda z} e^{j\frac{k}{2z}[x_0{}^2 + y_0{}^2]} \int_{-\infty}^{\infty} \int_{-\infty}^{\infty} P(x_1, y_1) f(x_1, y_1 : \omega) e^{j\frac{k}{2z}[x_1{}^2 + y_1{}^2]} e^{-j\frac{k}{2z}[x_0 x_1 + y_0 y_1]} dx_1 dy_1$$

$$\tag{2.11}$$

The integration is extended to infinity since there is no further contribution outside the aperture. We do this by inserting the Pupil function $P(x_1, y_1)$, which is the actual shape of the aperture. In the first form of Eq. (2.11), with a uniform illumination and a square aperture the integral separates into two one-dimensional integrals, which can be cast into the form of Fresnel integrals,

$$C(\alpha) = \int_0^\alpha \cos\frac{\pi t^2}{2}\, dt \qquad S(\alpha) = \int_0^\alpha \sin\frac{\pi t^2}{2}\, dt \tag{2.12}$$

which lead to the well-known Cornu spirals [1, p. 72].

2.3   The Fraunhofer approximation

Of more interest to us is the Fraunhofer approximation that uses the second form of Eq. (2.11), and covers the region

$$z >> \frac{k}{2} \left[x_1{}^2 + y_1{}^2 \right]_{max} \tag{2.13}$$

In this region the quadratic phase is much less than unity and has a minor contribution to the field at the receiver. In fact since we are mostly concerned with the energy in the field, we need only consider the integral

$$U(x_0, y_0 : z : \omega) = \frac{e^{jkz}}{\lambda z} \int_{-\infty}^{\infty} \int_{-\infty}^{\infty} f(x_1, y_1 : \omega) e^{-j\frac{2\pi}{\lambda z}[x_0 x_1 + y_0 y_1]} dx_1 dy_1 \tag{2.14}$$

Notice that this is the two-dimensional Fourier transform of the distribution in the aperture plane, $f(x_1, y_1 : \omega)$ evaluated at the frequency ω, multiplied by the factor $1/\lambda z$. We can also consider the spatial frequencies

$$f_x = \frac{x_1}{\lambda z} \equiv \frac{\alpha}{\lambda}$$

$$f_y = \frac{y_1}{\lambda z} \equiv \frac{\beta}{\lambda} \tag{2.15}$$

with (α, β) the directional cosines. This can be associated with the spatial domain as an angular spectrum, much as we associate frequency with the time domain [1, p. 55].

In Figure 2.2 we show the ranges of the two regions for apertures from 1 cm to 1 m, and wavelengths 0.5 μm to 10 μm.

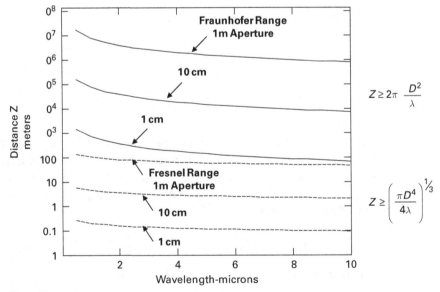

Figure 2.2

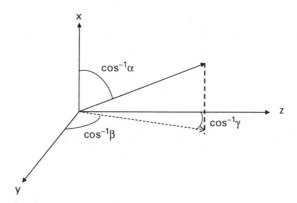

Figure 2.3 Directional cosines in vector space.

Example 2.1 To give physical meaning to the concept of spatial frequency [1, pp. 56–61] we consider k as a propagation vector $\mathbf{k}$ with magnitude $2\pi/\lambda$ and $k \cdot \mathbf{r} = \alpha x_1 + \beta y_1 + \gamma z$ with $(\alpha = x_1 /z = \lambda f_x, \beta = y_1 /z = \lambda f_y, \gamma = 1)$ the directional cosines and, with Figure 2.3, $\gamma = \sqrt{1-\alpha^2-\beta^2}$ One can develop diffraction theory using the assumption that the spatial frequencies of dominance in the diffraction pattern are the ones close to the optical axis, z, i.e., $\lambda^2 f_x^2 + \lambda^2 f_y^2 <<1$. In this case, one can approximate

$$\cos \gamma = \sqrt{1-\sin^2\gamma} \approx 1-1/2\sin^2\gamma \approx 1-1/2\left(\lambda^2 f_x^2 + \lambda^2 f_y^2\right)$$

and we obtain

$$U(x_0,y_0,z) \approx \iint\limits_{\alpha^2+\beta^2<1} A(\alpha/_\lambda,\beta/_\lambda,0)e^{j\frac{2\pi}{\lambda}\sqrt{1-\alpha^2-\beta^2}z}e^{j2\pi\left(\frac{\alpha x_0}{\lambda}+\frac{\beta y_0}{\lambda}\right)}d\left(\frac{\alpha}{\lambda}\right)d\left(\frac{\beta}{\lambda}\right) \qquad (2.16)$$

where

$$A(\alpha/_\lambda,\beta/_\lambda,0) = \int\limits_{-\infty}^{\infty}\int f(x_1,y_1,0)e^{-j2\pi\left(f_x x_0 + f_y y_0\right)}df_x\,df_y \qquad (2.17)$$

Equation (2.16) is known as the quadratic, or parabolic, approximation of the Huygens-Fresnel-Kirchoff integral. This approximation basically describes Fresnel diffraction by a Fresnel transformation.

Example 2.2 What are the validity ranges for the Fresnel and Fraunhofer transformations? These are given by Eqs. (2.10) and (2.13). In the former case, the region of validity is given by

$$\left[\frac{\pi}{4\lambda}\left((x_0-x_1)^2 + (y_0-y_1)^2\right)^2_{max}\right]^{1/3} << z \le \frac{k}{2}\left[x_1^2 + y_1^2\right]_{max}$$

$$z >> \left[\frac{\pi}{4\lambda}\left[(x_0-x_1)^2 + (y_0-y_1)^2\right]^2_{max}\right]^{1/3} \qquad (2.18)$$

The latter dictates that

$$z \gg \frac{k}{2} \left[x_1{}^2 + y_1{}^2 \right]_{\max} \tag{2.19}$$

Let us assume the wavelength of interest is 1.55 μm and we have our (maximum) aperture diameter, D, equal to 4 inches (= 10 cm). Then,

$$D^2 / \lambda = 6.450 \, \mathrm{km} \tag{2.20}$$

This is quite a distance between the source and observation plane. So, for short-range situations the diffraction pattern is dominated by Fresnel diffraction and for longer ranges it is dominated by Fraunhofer diffraction.

Example 2.3 Assume a plane wave is illuminating an object u, which is a rectangular aperture. Here, the observation plane is located a distance z away from the object. In this case, the resulting field amplitude in the observation plane will be dependent on the Fourier transform of the rectangular aperture, or:

$$U(x_0, y_0 : z : \omega) = \frac{e^{jkz} e^{j\frac{k}{2z}\left(x_0{}^2 + y_0{}^2\right)}}{j\lambda z} \int_{-\infty}^{\infty} e^{-j\frac{2\pi}{\lambda z}[x_0 x_{11}]} dx_1 \int_{-\infty}^{\infty} e^{-j\frac{2\pi}{\lambda z}[y_0 y_1]} dy_1$$

$$= \frac{e^{jkz} e^{j\frac{k}{2z}\left(x_0{}^2 + y_0{}^2\right)}}{j\lambda z} \cdot A \frac{\sin\left(\dfrac{2ax_0}{\lambda z}\right)}{\dfrac{2ax_0}{\lambda z}} \frac{\sin\left(\dfrac{2by_0}{\lambda z}\right)}{\dfrac{2by_0}{\lambda z}} \tag{2.21}$$

$2a$ is the length of the aperture in the x-direction and $2b$ is the length of the aperture in the y-direction. The $\sin x/x$ function has the property of being zero whenever x is an integer, except for $x = 0$, where $\sin 0/0 = 1$ (l'Hopital's rule).

The received intensity in the observation plane is then

$$|U(x_0, y_0 : z : \omega)|^2 = \frac{A^2}{\lambda^2 z^2} \left(\frac{\sin\left(\dfrac{2ax_0}{\lambda z}\right)}{\dfrac{2ax_0}{\lambda z}} \right)^2 \left(\frac{\sin\left(\dfrac{2by_0}{\lambda z}\right)}{\dfrac{2by_0}{\lambda z}} \right)^2 \tag{2.22}$$

Figure 2.4 shows a plot of $(\sin x/x)^2$. The width of the main lobe of $(\sin x/x)^2$ is 2, which implies that the widths of two $(\sin x/x)^2$ functions in Eq. (2.22) are equal to

$$\Delta x_0 = \lambda z / a \tag{2.23}$$

and

$$\Delta y_0 = \lambda z / b \tag{2.24}$$

respectively.

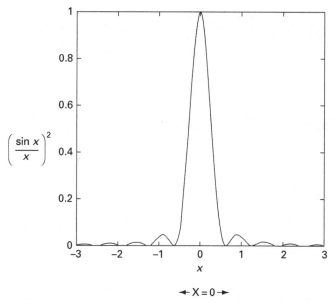

Figure 2.4 Plot of $(\sin x/x)^2$.

Example 2.4 Now assume a plane wave is illuminating an object, u, which is a circular aperture. Again, the observation plane is located a distance z away from the object. In this case, the resulting field amplitude in the observation plane is proportional to:

$$U(r_0 : \omega) = \frac{e^{jkz} e^{j\frac{k}{2z}r_0{}^2}}{j\lambda z} \int_0^a J_0\left(\frac{k\rho r_0}{z}\right)\rho d\rho$$

$$= \frac{e^{jkz} e^{j\frac{k}{2z}(x_0{}^2 + y_0{}^2)}}{j\lambda z} A \left(\frac{2J_1\left(\frac{kar_0}{z}\right)}{\frac{kar_0}{z}}\right) \qquad (2.25)$$

where r_0 is the radius of the aperture. In Eq. (2.25), $J_0(x)$ is the zeroth-order Bessel function and $J_1(x)$ is the first-order Bessel function. Squaring Eq. (2.25) with the proper normalizing term, we have

$$|U(r_0 : \omega)|^2 = \frac{A^2}{\lambda^2 z^2}\left(\frac{2J_1\left(\frac{kar_0}{z}\right)}{\frac{kar_0}{z}}\right)^2, \qquad (2.26)$$

which is the celebrated formula first derived in a different form by Airy. Figure 2.5 shows a plot of $[J_1(x)/x]^2$ as a function of x. This profile is known as the Airy Pattern. The first minima is located at $x = 1.22$, so the effective radius at that point is

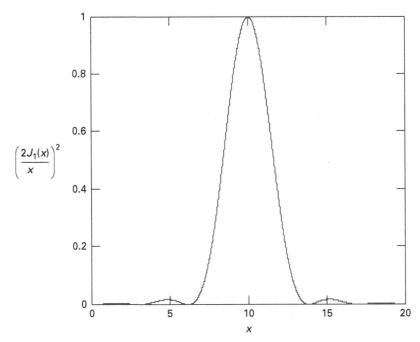

Figure 2.5 Plot of $(2J_1(x)/x)^2$.

$$\Delta w_0 = 1.22\,\lambda z / D \tag{2.27a}$$

$$\approx \lambda z / D \tag{2.27b}$$

Equation (2.26) is normally known as the beam-spreading term from diffraction-limited optics and its area form is used in typical link budgets. Calculating the area using Eq. (2.27b) gives

$$\pi\,\Delta w_0{}^2 \approx \pi\,[\lambda / D]^2 z^2 \tag{2.28}$$

or the solid angle subtended by the source in the observation plane is thus

$$\Omega_t \approx \frac{\lambda^2}{A_t} \tag{2.29}$$

[4, p. 8]. Recall that transmitter gain is defined as $G_t = 4\pi/\Omega_t$. Inserting Eq. (2.29) into this equation to yield

$$G_t = \frac{4\pi}{\Omega_t} = \frac{4\pi\,A_t}{\lambda^2} \tag{2.30}$$

Integrating Eq. (2.26) over 2π and radius w_0, we obtain

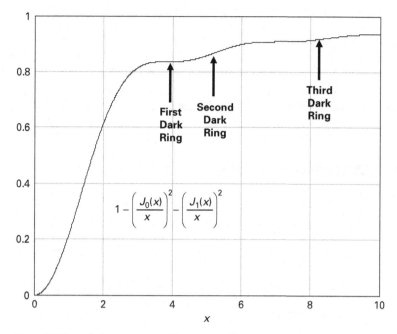

Figure 2.6 Cumulative energy profile versus radius.

$$L(w_0) = 1 - J_0^2\left(\frac{2\pi \, aw_0}{\lambda}\right) - J_1^2\left(\frac{2\pi \, aw_0}{\lambda}\right) \tag{2.31}$$

which was originally derived by Rayleigh. Figure 2.6 shows this function. It represents the total energy contained in the radius w_0. The fractional energy contained in the first, second and third rings is about 83%, 91% and 94%, respectively. From Figure 2.6 we see that in the first minima, or diffraction-limited spot, over 80% of the transmitted power is contained. The reader now can understand why this formula is "good enough" for link budget calculations.

The law of reciprocity [1, p. 41] states the equivalence of the aperture gain in either transmitting or receiving mode. Thus we see from example 1.3 that the range equation takes the form

$$P_r = \frac{P_t G_t A_r}{4\pi \, z^2} \cos \theta \tag{2.32}$$

where the receiver area is aligned at an angle θ with respect to the normal of the arriving ray direction from the transmitter. If we substitute for the gain of the transmitter, we get

$$P_r = \frac{P_t A_t A_r}{\lambda^2 z^2} \cos \theta \tag{2.33}$$

Alternatively, by the reciprocity theorem, we could have substituted for the area of the receiver to get

$$P_r = \frac{P_t G_t G_r \lambda^2}{(4\pi)^2 z^2} \cos\theta \qquad (2.34)$$

These are the two forms usually used in RF communications and radar system design. The quantity λ^2/A is commonly referred to as the diffraction-limited field of view of a system. In fact, when using the range equation, it is common practice to use the "effective area" of an antenna, defined as the equivalent circular area. When the effective area is not known, a gain pattern can be measured and used to determine the gain.

2.4   The mutual coherence function

If we define the covariance function as

$$E[f_{r_0}(\mathbf{r},\omega)f_{r'_0}{}^*(\mathbf{r}',\omega):z]$$

$$= \frac{1}{(\lambda z)^2}\underset{\Sigma\,\Sigma'}{\int\int} E[f_{r_1}(\mathbf{r_1},\omega)f_{r_1'}{}^*(\mathbf{r_1}',\omega)]e^{-j[(2\pi/\lambda^2 z^2)\mathbf{r_0}\cdot\mathbf{r_1}-\mathbf{r_0}'\cdot\mathbf{r_1}']}\,d\mathbf{r_1}d\mathbf{r_1}'$$

$$= \underset{\Sigma\,\Sigma'}{\int\int}\frac{\omega\varpi'}{(2\pi c)^2 r_{10}r'_{10}} R_{r_1 r_1'}(\mathbf{r_1},\mathbf{r_1}';\omega,\omega')e^{-j\left[\frac{\omega}{c}r_{10}-\frac{\omega'}{c}r'_{10}\right]}\,d\mathbf{r_1}d\mathbf{r_1}'$$

$$= R_{r_0}(\mathbf{r_0},\mathbf{r_0}';\omega:\omega') \qquad (2.35)$$

where $E[(\ )\}$ designates expectation, then the mutual coherence function (MCF) becomes [4, p. 33]

$$\Gamma(\mathbf{r_0},\mathbf{r_0}';t,t':) = \frac{1}{(2\pi)^2}\int_{-\infty}^{\infty} R_{r_0 r_0'}(\mathbf{r_0},\mathbf{r_0}';\omega:\omega')e^{-j(\omega t-\omega' t')}\,d\omega d\omega' \qquad (2.36)$$

and propagates in the same manner as does $f_{r_0}(\mathbf{r},t)$[5]. Equation (2.36) has the units of watts per unit area, or intensity (irradiance).

The MCF is discussed in detail in [2–5]. For our purpose we will use the engineering assumptions most useful in electro-optical system design. We will assume that the MCF is coherence separable into a spatial and a temporal component, that the spatial component is homogeneous, and that the temporal component is wide-sense stationary.

Coherence separable means that as far as the statistical characterization is concerned, time statistics are independent of the spatial statistics. This might not be the case if one was transmitting through a continuously deforming fog. In that case one would take snapshots of the fog at the Nyquist rate (of the fog) and treat each frame separately.

Homogeneous means that while a scene may change in space, the large variations behave independently of the local variations (or texture). Thus the assumption is that $R(r,r') \cong R(r,r+\rho) = I_s(r)R_s(\rho)$. Hence the scene is independently sensitive to both space variations and texture.

Wide-sense stationary has the same connotation in time as homogeneous has in space. That is we can separate the slow (diurnal) variations from the short-scale noise. Thus $R(t,t') \cong R(t,t+\tau) = R(\tau)$. Hence the scene is most sensitive to time differences.

When these assumptions are made, $R_{r_1 r_1'}(r_1,r_1';\omega,\omega')$ immediately separates into [4, pp. 37–41]

$$R_{r_1 r_1'}(\mathbf{r}_1, \mathbf{r}_1'; \omega) \equiv R_{r_1 r_1'}(\mathbf{r}_1, \mathbf{r}_1') R_s(\omega) \tag{2.37}$$

and using Eq. (2.36) and wide-sense stationarity, we integrate $R_s(\omega)$ to obtain $R_\tau(\tau)$. Using homogeneity, the remaining term in Eq. (2.35) can be written as

$$R_{\rho_0}\left(\frac{\boldsymbol{\rho}_0}{\lambda z}\right) = \frac{1}{(\lambda z)^2} \int_{\Sigma_1} \int_{\Sigma_1} I_s\left(\frac{\mathbf{r}_1}{\lambda z}\right) R_s\left(\frac{\boldsymbol{\rho}_1}{\lambda z}\right) e^{j\frac{2\pi[\mathbf{r}_1 \cdot \boldsymbol{\rho}_0 - \mathbf{r}_0 \cdot \boldsymbol{\rho}_1]}{\lambda^2 z^2}} d\mathbf{r}_1 d\boldsymbol{\rho}_1 \tag{2.38}$$

An incoherent source is defined as having random texture, or having $R(\rho_1) = \delta(\boldsymbol{\rho}_1)$. (Because the texture is generally rougher than an optical wavelength this is the general condition in optics, with the exception of optical quality surfaces that would transmit or reflect.) For this case the integral reduces to [5],

$$R_{\rho_0}\left(\frac{\boldsymbol{\rho}_0}{\lambda z}\right) = \frac{1}{\lambda z} \int_{\Sigma_1} I_s\left(\frac{\mathbf{r}_1}{\lambda z}\right) e^{j\frac{2\pi \boldsymbol{\rho}_0 \cdot \mathbf{r}_1}{\lambda^2 z^2}} d\mathbf{r}_1 \tag{2.39}$$

which we see is the Fourier transform of the large-scale intensity variation on the surface of the object and is the angular spectrum of the source. Equation (2.39) is called the Von Cittert–Zernike theorem [5].

The MCF becomes

$$\Gamma_{r_0 r_0'}(\tau, \rho_0) = R_\tau(\tau) R_{\rho_0}\left(\frac{\boldsymbol{\rho}_0}{\lambda z}\right) \tag{2.40}$$

where $R_\tau(\tau)$ is the temporal covariance function. Each has a Fourier transform, with $S_\tau(\omega)$ being the transform of $R_\tau(\tau)$, and $I_s(\mathbf{r}_1)$ being the transform of $R_s(\boldsymbol{\rho}_0)$, where $I_s(\mathbf{r}_1)$ is the large-scale intensity distribution over the source (the image).

Using the concept of spatial frequency, described earlier

$$f_x = \frac{x_1}{\lambda z} \Rightarrow df_x = \frac{dx_1}{\lambda z} : f_y = \frac{y_1}{\lambda z} \Rightarrow df_y = \frac{dy_1}{\lambda z}$$

$$\mathbf{f} = \frac{\mathbf{r}}{\lambda z} \Rightarrow d\mathbf{f} = \frac{d\mathbf{r}}{\lambda z} \tag{2.41}$$

we have

$$R_{\rho_0}\left(\frac{\boldsymbol{\rho}_0}{\lambda z}\right) = \frac{1}{\lambda z} \int_{-\infty}^{\infty} I_s(\mathbf{f}) e^{j2\pi \mathbf{f} \cdot \boldsymbol{\rho}_0} \frac{d\boldsymbol{\rho}_0}{\lambda z} \tag{2.42}$$

with

$$I_s(\mathbf{f}) = \int_{-\infty}^{\infty} R_s(\boldsymbol{\rho}_0) e^{-j2\pi \mathbf{f} \cdot \boldsymbol{\rho}_0} d\boldsymbol{\rho}_0 \tag{2.43}$$

and $\mathbf{f} = (f_x, f_y)$.

If we use the normalization suggested in [6, p. 87], we would have

$$\tilde{R}_s(0) = A_r \delta(0) \tag{2.44}$$

with

$$\tilde{R}_{\rho_0}\left(\frac{\boldsymbol{\rho}_0}{\lambda z}\right) = \frac{1}{\lambda z} \int_{-\infty}^{\infty} \left(\frac{A_r I_s(\mathbf{f})}{\int_{-\infty}^{\infty} I_s(\mathbf{f}) d\mathbf{f}}\right) e^{-j\frac{2\pi \mathbf{f} \cdot \boldsymbol{\rho}_0}{\lambda z}} d\boldsymbol{\rho}_0 = \frac{1}{\lambda z} \int_{-\infty}^{\infty} \tilde{I}_s(\mathbf{f}) e^{-j\frac{2\pi \mathbf{f} \cdot \boldsymbol{\rho}_0}{\lambda z}} d\boldsymbol{\rho}_0 \tag{2.45}$$

yielding

$$\tilde{R}_{\rho_0}(0) = A_r \tag{2.46}$$

Thus we see from Eq. (2.40) that

$$\tilde{R}_\tau(0)\tilde{R}_{\rho_0}(0) = \frac{\tilde{R}_{\rho_0}(0)}{2\pi} \int\limits_{-\infty}^{\infty} S_\tau(\omega)d\omega \tag{2.47}$$

and

$$\tilde{S}_\tau(\omega) = \frac{\tilde{R}_{\rho_0}(0)}{2\pi} S_\tau(\omega) \tag{2.48}$$

is the power spectrum, with $\tilde{R}_\tau(\tau)$ the inverse transform of $\tilde{S}_\tau(\omega)$.

Spatial coherence

The spatial covariance function is also called the spatial coherence function. By definition, $\tilde{R}_s(0) = 1$. This means that there is perfect correlation with a function and itself at the same point. The amount of correlation at the receiver is determined by the medium traversed. Any effect that decreases the coherence in the received field will tend to broaden or blur the image as described by the Fourier relationship [5].

Entropy

As a final point we note that the signal-to-noise ratio we have used before, as defined in Shannon's capacity equation, has not changed. Only now we have separated the temporal and spatial portions of the contributions. That is, the signal-to-noise ratio is properly defined as

$$\text{SNR} = \frac{P}{N} = \frac{\Gamma_f(0)}{\Gamma_n(0)} = \frac{\int\limits_{-\infty}^{\infty} \tilde{S}_\tau(\omega)d\omega}{\int\limits_{-\infty}^{\infty} \tilde{S}_n(\omega)d\omega} \frac{\int\limits_{-\infty}^{\infty} \tilde{I}_s(\mathbf{f})d\mathbf{f}}{\int\limits_{-\infty}^{\infty} \tilde{I}_n(\mathbf{f})d\mathbf{f}} \tag{2.49}$$

In those cases where we are looking at a frame of imaged data, we should properly define the signal-to-noise ratio as

$$\text{SNR} = \frac{P}{N} = \frac{\Gamma_f(0)}{\Gamma_n(0)} = \frac{\int\limits_{-\infty}^{\infty} \tilde{I}_s(\mathbf{f})d\mathbf{f}}{\int\limits_{-\infty}^{\infty} \tilde{I}_n(\mathbf{f})d\mathbf{f}} \tag{2.50}$$

since the integration time is constant. For those cases where the image is changing in time, then the image must be sampled at the Nyquist sampling rate associated with the time rate of change of the scene.

2.5   Linear transmission channels

If we refer back to the Fraunhofer approximation, we see that the field over the receiving plane is the inverse Fourier transform of the field over the transmitting plane. This is a linear, space-invariant operation by the channel to the transmitting beam, and should be multiplicative in the frequency domain. Therefore, if the channel effect $h(x_1, y_1 : \omega)$ is linear ($|h(x_1, y_1 : \omega)| \leq 1$), what the receiver receives will be the convolution

$$U(x_0, y_0 : \omega) = \frac{e^{jkz}e^{j\frac{k}{2z}(x_0^2+y_0^2)}}{j\lambda z} \int_{-\infty}^{\infty}\int_{-\infty}^{\infty}\int_{-\infty}^{\infty}\int_{-\infty}^{\infty} f(\varsigma, \xi : \omega)h((x_1-\varsigma),$$

$$(y_1-\xi)) : \omega)e^{j\frac{2\pi}{\lambda z}[x_0x_1+y_0y_1]}dx_1dy_1d\varsigma d\xi \tag{2.51}$$

of the field in space and the disturbance. This amounts to a blurring of each point in the field in space, and a stretching in time. Then, assuming the channel is coherence separable in time and space, we derive for the MCF [4,5],

$$\Gamma_{x_0y_0}(\tau, \mathbf{r}, \rho) = [R_S(\tau)\otimes M(\tau)][I_s(\mathbf{r})\otimes M(\mathbf{r})] \tag{2.52}$$

where now

$$S_0(\omega)M(\omega) = \int_{-\infty}^{\infty} [R_s(\tau)\otimes M(\tau)]e^{-j\omega\tau}d\tau \tag{2.53}$$

and

$$R_s(\boldsymbol{\rho}_0)M(\boldsymbol{\rho}_0) = \int_{-\infty}^{\infty} [I_s(\mathbf{f})\otimes M(\mathbf{f})]e^{-j2\pi\bar{\mathbf{f}}\cdot\boldsymbol{\rho}_0}d\mathbf{f} \tag{2.54}$$

with

$$M(\mathbf{r}_1) = [h(\mathbf{r}_1)\otimes h(\mathbf{r}_1)] \tag{2.55}$$

$M(\mathbf{r}_1)$ is referred to as the point source coherence function. The significance of this can be seen if we assume a point source, represented as

$$I\{\mathbf{r}_1\} = I_1\delta(\mathbf{r}_1) \tag{2.56}$$

yielding

$$I(\mathbf{r}_1)\otimes M_{r_{01}}(\mathbf{r}_1) = I_1M_{r_{01}}(\mathbf{r}_1) : \quad \leq I_1 \quad \forall \mathbf{r}_1$$

$$\text{Since} \quad M_{r_{01}}(\mathbf{r}_1) \leq 1 \quad \forall \mathbf{r}_1 \tag{2.57}$$

Thus we see that any non-unity value for $M(\mathbf{r}_1)$ results in blurring of the image $I_s(\mathbf{r}_1)$, or filtering in the spatial frequency domain. When the spatial coherence function is rotationally invariant, it is called isotropic. For the time characterization of random channels, refer to [7]. It is often the case where the point spread coherence function has a random component to it which can also cause a filtering effect. We have used both notations $\mathbf{r}$ and $\mathbf{f}$ from Eq. (2.41) interchangeably.

2.6 Thin lens receiver optics

It can be shown [1, p. 98] that if we construct a lens with spherical front and back surfaces, and radii of curvature L_1, L_2 (Figure 2.7), then the same Fresnel approximation

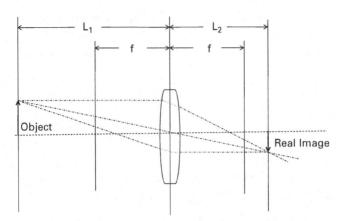

Figure 2.7 Ray optics description of a focusing lens.

can be made to create an equation identical with the Fraunhofer approximation, in which the field in the detector plane becomes

$$U(r_d : \omega) = \frac{e^{jkf_c}}{j\lambda f_c} e^{j(k/2f_c)r_d^2} \int_{-\infty}^{\infty} \int_{-\infty}^{\infty} f_{r_0}(\mathbf{r_0} : \omega) e^{-j(k/f_c)[\mathbf{r_d \cdot r_0}]} d\mathbf{r_0} \qquad (2.58)$$

where we have assumed the thin lens theory to cancel the quadratic term inside the integral. This holds for

$$\frac{1}{L_1} + \frac{1}{L_2} - \frac{1}{f_c} = 0 \qquad (2.59)$$

f_c is called the focal length of the lens, which for $L_1 \gg f_c$, approaches L_2. Equation (2.59) is known as the lens law [1]. If A_d is the area of the detector, then A_d/f_c^2 is the field of view of the receiver. If the detector has many pixels, each with area A_p, then A_p/f_c^2 is the pixel field of view, sometimes called the instantaneous field of view. Since λ^2/A_r would be the minimum resolvable spot, we have that

$$A_d/_{fc} \geq \lambda^2/_{A_r} \qquad (2.60)$$

From examples 2.1 and 2.2, we see that this is also approximately

$$X_p/_{f_c} \geq \lambda/_D \quad or$$

$$X_p/_{\lambda} \geq f_c/_D = F \qquad (2.61)$$

where X_p is the dimension of a pixel, D is the dimension of the lens, and F is called the F-number of the lens. When the equality holds, the system is called diffraction limited.

The spatial intensity at the detector becomes

$$I_d(r_d) = \frac{1}{2\pi} \int_{-\infty}^{\infty} |U(r_d : \omega)|^2 d\omega \qquad (2.62)$$

Example 2.5 Suppose we have a circular lens at the receiver with diameter D. The field at the detector now includes a pupil function for the lens and can be written as

$$U(r_d : \omega) = \frac{e^{jkf_c}}{j\lambda f_c} e^{j\left(k/2f_c\right)r_d^2} \int_{-\infty}^{\infty}\int_{-\infty}^{\infty} f_{r_0}(\mathbf{r_0} : \omega)h(\mathbf{r_0})P(\mathbf{r_0})e^{-j\left(k/f_c\right)[\mathbf{r_d}\cdot\mathbf{r_0}\,1]}d\mathbf{r_0} \quad (2.63)$$

The spatial intensity at the detector now becomes

$$I_d(r_d) = \int_{-\infty}^{\infty} I_s(\mathbf{f})d\mathbf{f} \int_{-\infty}^{\infty} M(\lambda f_c\zeta)\Theta(\lambda f_c\zeta)e^{-j2\pi\zeta\cdot(\lambda f_c\mathbf{f}+\mathbf{r_d})}d\zeta \quad (2.64)$$

where

$$\Theta(\lambda f_c\zeta) = \int_{-\infty}^{\infty} P(\lambda f_c\mathbf{f_0})P(\lambda f_c(\mathbf{f_0}+\zeta)d\zeta \quad (2.65)$$

is the OTF or optical transfer function of the lens. This is merely the convolution of the pupil function with itself. If there is no coherence distortion, $M(\lambda f_c\zeta)\Theta(\lambda f_c\zeta) = 1$, and $I_d(r_d)$ reduces to

$$I_d(\mathbf{r_d}) = \int_{-\infty}^{\infty} I_s(\mathbf{f})d\mathbf{f} \int_{-\infty}^{\infty} e^{-j2\pi\zeta\cdot(\lambda f_c\mathbf{f}+\mathbf{r_d})}d\zeta$$

$$= \int_{-\infty}^{\infty} I_s(\mathbf{f})\delta(\lambda f_c\mathbf{f}+\mathbf{r_d})d\mathbf{f}$$

$$= \left(\frac{1}{\lambda f_c}\right)I_s\left(-\frac{\mathbf{r_d}}{M}\right) \quad (2.66)$$

This is merely the original image, reversed and magnified by $M = L_1 / L_2$. In fact, we will always have a finite aperture, usually circular. For this case the OTF becomes [1, p. 146]

$$\Theta(\lambda f_c\zeta) = \begin{cases} \frac{2}{\pi}\left[\cos^{-1}\left(\frac{\zeta}{d}\right) - \frac{\zeta}{d}\sqrt{1-\left(\frac{\zeta}{d}\right)^2}\right] : & \zeta \le d = \frac{D}{\lambda f_c} \\ 0 & \text{elsewhere} \end{cases} \quad (2.67)$$

Example 2.6 Example OTF calculation with aberrations.

Following Goodman [1, pp. 123–125], one can calculate the OTF of a simple lens system with focus error. In this example, a rectangular aperture will be assumed to simplify the analysis.

If we have an image slightly out of focus, then Eq. (2.59) will be rewritten as

$$\frac{1}{L_1} + \frac{1}{L_2} - \frac{1}{f_c} = \varepsilon \quad (2.68)$$

The aberration function is then written as

$$W(x,y) = \varepsilon(x^2 + y^2)/2 \quad (2.69)$$

For a square aperture of width $2a$, the maximum phase error at the edge of the aperture (along the x- or y-axis) is given by

$$w = \varepsilon(2a)^2/8 \quad (2.70)$$

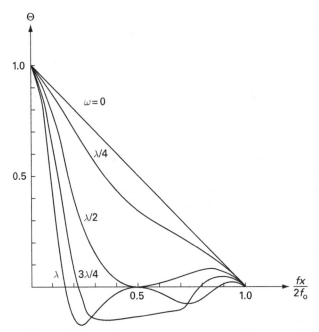

Figure 2.8 Cross section of the OTF for a focusing error and a square aperture [8, p. 125].

The number w is a convenient indication of the severity of the aberration, error of focus in this case. Given this situation, Goodman showed that the above aberration yielded the following OTF

$$\Theta(f_x,f_y) = \Lambda\left(\frac{f_x}{2f_0}\right).\Lambda\left(\frac{f_y}{2f_0}\right).\frac{\sin\left[\frac{8\pi w}{\lambda}\left(\frac{f_x}{2f_0}\right)\left(1-\frac{|f_x|}{2f_0}\right)\right]}{\left[\frac{8\pi w}{\lambda}\left(\frac{f_x}{2f_0}\right)\left(1-\frac{|f_x|}{2f_0}\right)\right]}.\frac{\sin\left[\frac{8\pi w}{\lambda}\left(\frac{f_y}{2f_0}\right)\left(1-\frac{|f_y|}{2f_0}\right)\right]}{\left[\frac{8\pi w}{\lambda}\left(\frac{f_y}{2f_0}\right)\left(1-\frac{|f_y|}{2f_0}\right)\right]}$$

$$(2.71)$$

where

$$\Lambda(x) = 1-|x| : \quad for \quad |x| \leq 1$$

$$= 0 \quad otherwise \qquad (2.72)$$

Figure 2.8 plots Eq. (2.71) for several fraction values of defocusing [8, p. 125]. In this figure, a value for w of zero equates to diffraction-limited imaging, the best one can obtain. For non-zero values of w, the OTF degrades and experiences contrast reversals for values of w greater than half a wavelength.

Example 2.7 Strehl ratio

When aberrations are induced by a lens, or the turbulence in an incoherent optical imaging system, the transmitted irradiance is reduced by a factor commonly called the

Strehl ratio (SR) [2, p. 462; 8, p. 139]. The modern definition of the Strehl ratio is the ratio of the observed peak intensity at the detection plane of a telescope or other imaging system from a point source compared to the theoretical maximum peak intensity of a perfect imaging system working at the diffraction limit. In terms of the above example, it is defined as the ratio of the light intensity at the maximum of the point spread function (impulse response) of the optical system with aberrations to the maximum of the point spread function of the same system without aberrations. Alternatively when aberrations are severe ($w \gg \lambda$), the Strehl Ratio is equal to the normalized volume under the OTF of the aberrated optical systems, or more specifically,

$$SR = \frac{\iint_{-\infty}^{\infty} \Theta(f_x, f_y)|_{with\ aberrations}\ df_x\ df_y}{\iint_{-\infty}^{\infty} \Theta(f_x, f_y)|_{without\ aberrations}\ df_x\ df_y}, \tag{2.73}$$

[8, p. 139].

2.7 Geometrical analysis for diffuse illumination into an optical imaging system

When diffuse light sources are being imaged into an optical system, it is simpler to use geometrical optics to calculate the received power at the detector/focal plane. Let us calculate the received power by a lens from a Lambertian surface, a typical extended source type in many optical problems [8, pp. 207–211]. Figure 2.9 illustrates the basic

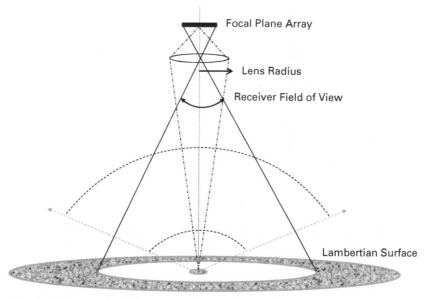

Focal Plane Array

Lens Radius

Receiver Field of View

Lambertian Surface

Figure 2.9 Power received from a Lambertian source.

geometry of this situation. In this figure, the radius of the circular lens is a, the radius of the source is r_0, the solid angle of the entire focal plan array is Ω_{FOV}, the radius of the source that images into an individual detector array element is r_e, and the solid angle of an individual focal plane array element is Ω_{IFOV}.

A Lambertian surface is one where the extended surface emitting optical radiation follows Lamberts' Law in its angular distribution, or

$$J_L = J_0 \cos\theta \qquad (2.74)$$

where J_L is the radiant intensity of a small incremental area of the source in the direction of angle θ from the surface normal and J_0 is the radiant emittance of a small incremental area of the source in the direction of the surface normal [8, p. 207].

Consider an elemental area, dA, which radiates power into an elemental solid angle $d\Omega$ as shown in Figure 2.10. The spectral radiance, N_λ, is related to the spectral radiant emittance by the relationship (Chapter 1, Eq. (1.7)).

$$W_\lambda = \iint\limits_{\Omega_{Hemisphere}} N_\lambda \cos\theta d\Omega \qquad (2.75)$$

where $d\Omega = \sin\theta\, d\Omega_{Hemisphere}$.

The total received spectral power in the hemisphere from a source area A is then given by

$$P_\lambda = \iint\limits_{A} \iint\limits_{\Omega_{Hemisphere}} N_\lambda \cos\theta d\Omega dA$$

$$= \pi N_\lambda A = W_\lambda A \qquad (2.76)$$

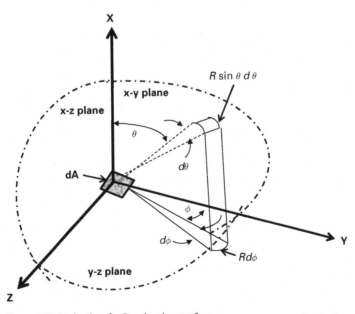

Figure 2.10 Projection for Lambertian surface.

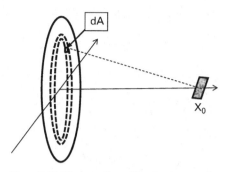

Figure 2.11 Elemental are illuminating a point x_0.

Thus, the spectral radiant emittance and spectral radiance are related by a constant for Lambertian surfaces. The total power is then

$$P = \int P_\lambda (\lambda) d\lambda \approx \pi N_\lambda A \, \Delta\lambda \qquad (2.77)$$

for most practical situations. Here, $\Delta\lambda$ is the receiver optical spectrum bandwidth.

We now will calculate the received power incident on an individual focal plane array element. Let the Lambertian source be positioned normal to the z axis with radiance N_λ and area A. Define $x_0 = (0, 0, z_0)$ to be a point located a distance z_0 from the source on the normal axis [8]. Figure 2.11 shows this situation graphically. Since the distance from dA to x_0 is $z_0 \sec \theta$, and $dA = 2\pi z_0 \tan \theta z_0 \sec^2 \theta d\theta$, the incremental radiance at x_0 is

$$dH_\lambda = N_\lambda dA \cos \theta \left[\frac{\cos^3 \theta}{z_0{}^2} \right] = 2\pi N_\lambda \sin \theta \cos \theta \, d\theta \qquad (2.78)$$

[8]. Integrating around the source, the irradiance for the entire source is given by

$$H_\lambda = 2\pi \int_0^{\theta_1} N_\lambda \sin \theta \cos \theta \, d\theta$$

$$= \frac{\pi N_\lambda r_e{}^2}{(r_e^2 + z_0^2)} \qquad (2.79)$$

where $\theta_1 = \tan^{-1} \left(\frac{r_e}{z_0} \right)$.

If our point of illumination does not lie on the z-axis, the irradiance is subject to the "cosine to the fourth law". This comes from a $\cos \theta$ at the object, a $\cos \theta$ at the receiver and a $1/\sec^2 \theta$ in between. In other words, the irradiance at a point x_0' making an angle θ' with the z-axis can be written as

$$H_\lambda'(\theta') = H_\lambda \cos^4 \theta \qquad (2.80)$$

with H_λ being the irradiance calculated above.

Taking a ring at the aperture in order to sum, we have

$$dP_{T\lambda}' = 2\pi H_\lambda \cos^4 \theta \, \rho \, d\rho \qquad (2.81)$$

which leads to the expression for the power at the aperture

$$P_{T\lambda}{}'=2\pi \int_0^{\theta_0{}'} H_\lambda \cos\theta' \sin\theta' d\theta'$$

$$=\frac{(\pi a^2)z_0^2 N_\lambda(\pi r_e^2)\Delta\lambda}{(a^2+z_0^2)(r_e^2+z_0^2)}$$

$$=N_\lambda A_{rec}\Omega_{IFOV}\Delta\lambda \qquad \text{for } a \ll z_0 \qquad (2.82)$$

with

$$\Omega_{IFOV} \equiv DetectorSolidAngle = \frac{\pi r_e^2}{(r_e^2 + z_0^2)} \qquad (2.83)$$

in agreement with Eq. (1.11). Equation (2.80) is the engineering equation used to determine whether there is enough signal in the array element to overcome the system and received noise.

For the total power received by the entire detector array, we have

$$P_{T\lambda}{}'=2\pi \int_0^{\theta_0{}'} H_\lambda \cos\theta' \sin\theta' d\theta'$$

$$=\frac{(\pi a^2)z_0^2 N_\lambda(\pi r_o^2)\Delta\lambda}{(a^2+z_0^2)(r_o^2+z_0^2)}$$

$$=N_\lambda A_{rec}\Omega_{FOV}\Delta\lambda \qquad \text{for } a \ll z_0 \qquad (2.84)$$

with

$$\Omega_{FOV} \equiv \text{detector array solid angle}$$

$$=\frac{\pi r_0^2}{(r_0^2 + z_0^2)} \qquad (2.85)$$

Example 2.8 Cosine to the fourth

The "cosine to the fourth law" is a very important first-order effect in the design of any optical system. It is more than the keystone effect, also known as the tombstone effect, which is caused by attempting to project an image onto a surface at an angle, as with a projector not quite centered onto the screen it is projecting on. It is a distortion of the image dimensions. The cosine to the fourth effect includes the keystone effect, plus three other angular effects, which account for the irradiance reduction at an axis point located an angle θ off the optical center line. In this example, we will provide the derivation of the "cosine to the fourth law" per Smith [8, pp. 144–145].

Figure 2.12 shows the relationship between the exit pupil and the image plane for a point A on the optical axis and an off-axis point H. The irradiance received at the latter point is proportional to the solid angle that the exit pupil subtends from the point. Let us now look at the details.

The solid angle subtended from point A is the area of the exit pupil, divided by the square of the distance OA, the slant distance from the center of the exit pupil to the point H. From H, the solid angle is the projected area of the pupil, divided by the square

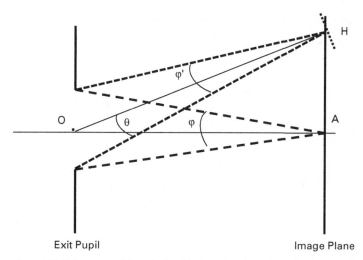

Exit Pupil   Image Plane

Figure 2.12 Schematic of the relationship between the exit pupil and the image plane used to derive the cosine to the fourth law.

of the distance OH. Since OH is greater than OA, the distance from the exit pupil to the image plane by a factor of $1 / \cos \theta$, this increased distance reduces the irradiance by a factor of $\cos^2 \theta$. Here, θ is the angle between point A and point H relative to the exit pupil. In addition, the exit pupil is viewed obliquely from the point H, and its projected area is reduced by a factor proportional to $\cos \theta$ when OH is large compared to the size of the exit pupil. The result is that the irradiance at H is reduced by $\cos^3 \theta$.

Finally, this last result only is true for illumination on a plane normal to the line OH, indicated by the small dashed line at H in Figure 2.12. However, we want the irradiance in the plane AH, which reduces the irradiance by another factor of $\cos \theta$. This implies that the irradiance at H in the plane AH is equal to the irradiance at point A times $\cos^4 \theta$. Figure 2.13 depicts cosine to the fourth as a function of angle. The importance of this effect comes clear for wide-angle imaging systems since the irradiance reduction at 30° (0.524 radians), 45° (0.785 radians), 60° (1.047 radians) is 0.56, 0.25 and 0.06, respectively, which are quite significant values.

As a final note, the above is a good approximation when OH is large compared to the size of the exit pupil. For large angles where this is not true, P. Foote [9], in the *Bulletin of the Bureau of Standards*, gave a more accurate equation for the irradiance at point H; namely,

$$H = \frac{\pi N}{2} \left[1 - \frac{(1 + \tan^2 \varphi - \tan^2 \theta)}{(\tan^4 \varphi + 2 \tan^2 \varphi (1 - \tan^2 \theta) + 1 / \cos^4 \theta)^{1/2}} \right] \qquad (2.86)$$

For extreme cases, Smith showed that Eq. (2.86) can provide a 42% larger irradiance that the cosine to the fourth [8, p. 215]. However, for most practical applications, cosine to the fourth is sufficient.

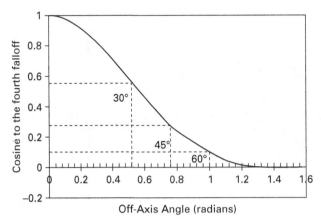

Figure 2.13 Cosine to the fourth as a function of angle.

2.8   Summary

In this chapter we have taken the reader quickly from the wave equation, through the Fresnel-Kirkoff development of point-to-point electromagnetic transport, to the application of a transmitting aperture transmitting through a linear channel, to the reception of a thin optical lens, focused onto a finite detector. We saw that due to the wave nature of the field, diffraction effects occur, which limit resolution in angle, and establish minimum angles and resolution. Our intention was to show the reader how the different elements of real systems play in the transport of electromagnetic energy. We have also extended the Capacity theory to include spatial effects. In later chapters we will address aspects of this transfer problem.

References

1. J. W. Goodman. *Introduction to Fourier Optics*, 3rd edn. Roberts and Company, Englewood. CO (2004).
2. L. C. Andrews and R. L Phillips. *Laser Propagation through Random Media*, 2nd edn. SPIE Press, Bellingham, WA (2005).
3. M. Born and E. Wolf. *Principles of Optics: Electromagnetic Theory of Propagation, Interference and Diffraction of Light*, 7th edn. Cambridge University Press, Cambridge (1999).
4. S. Karp, R. M. Gagliardi, S. E. Moran and L. B. Stotts. *Optical Channels*. Plenum Press, New York (1988).
5. A. Papoulis. *Systems and Transforms with Applications to Optics*. McGraw-Hill, New York (1968) p. 383.
6. R. M. Gagliardi and S. Karp. *Optical Communications*. Wiley Interscience, New York (1976).
7. P. A. Bello. Characterization of randomly time-variant linear channels. Section V, *IEEE Trans. Comm. Systems*, CS-11 (1963), p. 360.
8. W. J. Smith. *Modern Optical Engineering*. McGraw Hill, New York (1990).
9. P. Foote. *The Bulletin of the Bureau of Standards*, **12** (1915), p. 583.

3 Photo-detection of electromagnetic radiation

In Chapters 1 and 2 we showed that we could quantitatively trace an electromagnetic field from the source through a channel, and the receiving optics to the detector. This path was summarized in Eqs. (2.58), (2.62), (2.63) and (2.66). In particular we see that we can trace the path from the source to an intensity incident on the detector in the receiver focal plane. In this chapter we will characterize the response of a photo-detector to an incident optical field, along with the characteristics of the resulting current and how it reflects the properties of the incident radiation.

3.1 The photo-detector

When intensity is incident on a quantum detector, the result is a release of electrons. The behavior of the electrons has been discussed in [1, p. 51], and is characterized by a distribution of electron counts. This characterization can be visualized as a two-dimensional spatial field over an area A_d, evolving in time T. Figure 3.1 shows this in pictorial form, with $A_d T = V$, being the "volume" of integration. The spatial variable is $\mathbf{r}$, and the time variable is t. Since we are considering the interaction between a field and a molecule, we are interested in the infinitesimal volume $d\mathbf{r}dt = dv$. The release of an electron is therefore

$$\left| \text{Probability of an electron emitted from } \Delta v_i \text{ at point } v_i \right| = \alpha I(v_i)\Delta v_i$$

$$\left| \text{Probability of no electron being emitted} \right| = 1 - \alpha I(v_i)\Delta v_i \tag{3.1}$$

where α is a constant of proportionality. In the limit as we let $\Delta v \to 0$ we can derive a counting distribution for the released electrons to be $P_k(k)$ where

$$P_k(k) = \frac{(m_v)^k}{k!} e^{-m_v} \tag{3.2}$$

and k is the total number of electrons released in the volume V. This is the well-known Poisson distribution. m_v assumes the form

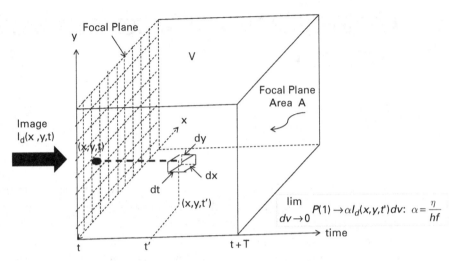

Figure 3.1 Geometry used for deriving Poisson probability density in three dimensions.

$$m_v = \alpha \int\limits_{A_d} \int\limits_t^{t+T} I_d(\rho, \mathbf{r}) d\rho d\mathbf{r} \qquad (3.3)$$

Dimensionally we see that $I(\mathbf{v})$ has the dimensions of power per unit area, multiplied by area to yield power, and then by time to yield energy. Since, by the laws of thermo-dynamics, this conversion is always less than or equal to unity, there is always a "quantum efficiency", η, associated with this process. It is also convenient to introduce the variable

$$n(t, \mathbf{r}) \equiv \alpha I(t, \mathbf{r}) \qquad (3.4)$$

as the "rate" of electron release over the elementary volume $d\mathbf{r}dt = d\mathbf{v}$. When an area is specified, we can further reduce this to a rate per unit time $n(t)$, where

$$n(t) \equiv \alpha \int\limits_{A_d} I(t, \mathbf{r}) d\mathbf{r} \qquad (3.5)$$

This is true whenever spatial effects can be ignored. The time averaged rate becomes

$$\bar{n}_T \equiv \frac{1}{T} \int\limits_t^{t+T} n(t) dt \qquad (3.6)$$

We see [1, p. 56] that m_v is also the average value of the parameter k, the number of counts released in the volume V. Thus

$$E_k[k] = m_v = \alpha \int\limits_{A_d} \int\limits_t^{t+T} I_d(\rho, \mathbf{r}) d\rho d\mathbf{r} \qquad (3.7)$$

From the quantum effect, first introduced by Einstein [3, Vol. II, p. 89], it takes an amount of energy equal to hf at the frequency f to release an electron. Thus we can also consider the quantum efficiency to be equal to the probability that an electron will be released, given that an amount of energy hf was incident on the detector. From this we can deduce that the average number of electrons released requires an average amount of absorbed energy equal to

$$E_k[k]hf = \eta \int_{A_d} \int_t^{t+T} I_d(\rho, \mathbf{r}) d\rho d\mathbf{r} \tag{3.8}$$

and the parameter α can now be evaluated as

$$\alpha = \frac{\eta}{hf} \tag{3.9}$$

We also know from the first Law of Thermodynamics that energy cannot be created, hence $\eta \leq 1$. Other properties of this Poisson variable have been identified in [1, Ch. 2].

3.2   Shot noise processes

When an electron is released from the quantum detector, it flows through a circuit as a current whose integral is equal to the electron charge e. That is

$$i = \frac{dq}{dt} \tag{3.10}$$

with

$$\int i(t)dt = \int \frac{dq}{dt} dt = \int dq = e \tag{3.11}$$

The characteristics of the circuit will determine the shape of $i(t)$, but the integral will always be the same, e. If we define the shot noise process as the current produced by the ensemble of electrons released in the time interval T, over an area A, then we can represent this ensemble as

$$x(\mathbf{r}, t) = \sum_{j=1}^{k} h[(\mathbf{r}-\mathbf{r_j}); (t-t_j)] \quad : \quad \mathbf{r_j}, t_j \in [A; (t, t+T)] \tag{3.12}$$

and use the fact that the events are independent and identically distributed in the volume with probability density [1, p. 110] (Appendix A),

$$p_V(\mathbf{r_j}, t_j) = \frac{n(\mathbf{r_j}, t_j)}{\int_A \int_t^{t+T} n(\mathbf{r}, t) d\mathbf{r} dt} \tag{3.13}$$

We next integrate over the area to reduce this to the time process with probability density over the $(t, t+T)$ interval

$$p_T(t_j) = \frac{n(t_j)}{\int_t^{t+T} n(t)dt} \tag{3.14}$$

Although we assume no spatial effects, these can be introduced later. We can compute the mean and variance of the time process $x(t)$. We then have for the mean

$$E[x(t)] = \int_t^{t+T} h(t-z)n(z)dz \tag{3.15}$$

Thus we see that the current reproduces the incident intensity, filtered by $h(z)$, which is the impulse response of the photo-detector. If the duration of $h(z)$, τ_e, is short compared to the variations in $n(z)$, it can be approximated by $h(z) \approx e\delta(z)$, hence

$$E[x(t)] \approx \int_t^{t+T} e\delta(t-z)n(z)dz = en(t) \tag{3.16}$$

Since $n(t) \equiv \alpha \int_{A_d} I(t,r)dr \geq 0$, we can represent Eq. (3.14) as

$$p_T(t_j) = \frac{n(t_j)}{\int_t^{t+T} n(t)dt} = \frac{\bar{n}_T(1+m(t_i))}{\bar{n}_T T}$$

$$= \frac{(1+m(t_j))}{T} \quad : \quad m(t) \geq -1 \tag{3.17}$$

with

$$\int_t^{t+T} m(t)dt = 0 \tag{3.18}$$

Similarly, the variance of the process can be computed to be

$$\text{Var}[x(t)] = \int_t^{t+T} h^2(t-z)n(z)dz \tag{3.19}$$

For the short-duration pulse, we can approximate $h(z)$ by

$$h(z) \approx \frac{e}{\tau_e} \quad : \quad 0 \leq z \leq \tau_e \tag{3.20}$$

and compute the variance to be

$$\text{Var}[x(t)] \approx \int_t^{t+\tau_e} \left(\frac{e}{\tau_e}\right)^2 n(z)dz = \left(\frac{e}{\tau_e}\right)^2 k(t, t+\tau_e) \tag{3.21}$$

where $k(t, t+\tau_e) = \tau_e n(t)$ is the count in the interval $(t, t+\tau_e)$. Thus for a wideband shot noise process, there is an inherent signal-to-noise ratio of

$$\text{SNR} = \frac{(en(t))^2}{e^2 n(t)/\tau_e} = n(t)\tau_e = k(t, t + \tau_e) \tag{3.22}$$

in other words, the inherent capacity is independent of additional noise. For $n(t)\tau_e \gg 1$, this distribution can be shown to approach the Gaussian distribution [1, p. 111]. In fact in this limit the process $x(t)$ approaches the rate $n(t)$ with a variance that is Gaussian in nature. This implies that the variance of the "shot noise" appears Gaussian for large values of intensity. If we identify a signal bandwidth $\tau_e = 1/2\, B_s$, the SNR becomes ($I(t)A_d = P(t)$, the detected power)

$$\text{SNR} = \frac{\eta P(t)}{2hfB_s} \tag{3.23}$$

Notice that hf replaces kT as the apparent noise level. This quantization limit on SNR is referred to as the "quantum-limited" signal-to-noise ratio.

3.3 Power spectral density of shot noise

To compute the power spectral density (PSD) of the shot noise process $x(t)$ [1, p. 119], we first take the Fourier transform

$$X_T(\omega) = \int_{-T}^{T} \sum_{j=1}^{k} h(t - t_j) e^{-j\omega t} dt$$

$$= \sum_{j=1}^{k} e^{-j\omega t_j} H_T(\omega) \tag{3.24}$$

The PSD is defined as

$$S_x(\omega) \equiv \lim T \to \infty \frac{1}{2T} E\left[|X_T(\omega)|^2\right]$$

$$= e^2 |H(\omega)|^2 \left[\bar{n} + \bar{n}^2 \left[2\pi\delta(\omega) + |S_m(\omega)|^2\right]\right] \tag{3.25}$$

where

$$\bar{n} = \lim T \to \infty \bar{n}_T$$

$$S_m(\omega) = \lim T \to \infty \left[\frac{1}{2T} E_m\left[|M_T(\omega)|^2\right]\right] \tag{3.26}$$

Here the function $m(t)$ (the modulation index) is also allowed to be a random process, having a PSD of its own. Thus an additional average over this process is required. This is also referred to as a doubly stochastic shot noise process. By using Parseval's theorem, we have

$$\frac{1}{2\pi} \int\limits_{-\infty}^{\infty} S_m(\omega)d\omega = \lim T \to \infty \frac{1}{2T} \int\limits_{-T}^{T} |m(t)|^2 dt = m^2 \tag{3.27}$$

We have defined m^2 as the time average modulation index (depth of modulation), but $m(t)$ need only be greater than -1 at any time t. Notice also that there is a white noise level $\bar{n}$ which is filtered by the spectrum of the pulse shape $H(\omega)$. So again, in the absence of any additive noise, we have for the signal-to-noise ratio (using only the AC portion of the spectrum)

$$\text{SNR} = \frac{\dfrac{(\bar{n}e)^2}{2\pi} \int\limits_{-\infty}^{\infty} |H(\omega)|^2 S_m(\omega)d\omega}{\dfrac{\bar{n}e^2}{2\pi} \int\limits_{-\infty}^{\infty} |H(\omega)|^2 d\omega} = \frac{\bar{n}m^2}{2B_m} \tag{3.28}$$

Since we have defined $\bar{n}$ as the average electron rate, we have

$$\bar{n} \equiv \lim T \to \infty \left[\frac{1}{T} \int\limits_{t}^{t+T} n(t)dt \right] = \frac{1}{T} \int\limits_{t}^{t+T} dt \frac{\eta}{hf} \int\limits_{A_d} I(t,r)dr$$

$$= \int\limits_{t}^{t+T} dt \frac{\eta}{hf} \left[\frac{\int\limits_{A_d} I(t,r)dr}{T} \right] = \frac{\eta}{hf} \int\limits_{t}^{t+T} \frac{P(t)}{T} dt = \frac{\eta \bar{P}_m}{hf} \quad : \quad \bar{P}_m \equiv m^2 \bar{P} \tag{3.29}$$

and

$$\text{SNR} = \frac{\dfrac{(\bar{n}e)^2}{2\pi} \int\limits_{-\infty}^{\infty} |H(\omega)|^2 S_m(\omega)d\omega}{\dfrac{\bar{n}e^2}{2\pi} \int\limits_{-\infty}^{\infty} |H(\omega)|^2 d\omega} = \frac{m^2 \bar{n}}{2B_m} = \frac{\eta m^2 \bar{P}}{2hfB_m} \tag{3.30}$$

in the Shannon definition of SNR. Since only the AC portion of the signal contributes to capacity, we have included the average value of the modulation index m in the definition. This represents the fraction of the intensity that is useful signal (modulation index) and is defined as

$$m^2 \equiv \frac{\int\limits_{t}^{t+T} |m(t)|^2 dt}{T} \tag{3.31}$$

It is also important to point out that since the detector is an energy detector, the current is proportional to the energy as a function of time. Therefore the signal to be transferred is intensity-modulated and not amplitude- or phase-modulated, as in radio frequency communication systems. It is possible to modulate the amplitude, but this will be covered later in sections on coherent systems. Technically speaking, we have only defined the current process. To make this power we assume a 1 ohm load.

3.4 A general solution to the counting distribution

Recall that the mean value of the count distribution was

$$m_v = \alpha \int_{A_d} \int_t^{t+T} I_d(\rho, \mathbf{r}) d\rho d\mathbf{r} \tag{3.32}$$

Since $I_d(\rho, \mathbf{r})$ was the intensity of the field when it was focused on the detector, it was also proportional to

$$m_v = \frac{\eta}{hf} \int_{A_d} \int_t^{t+T} E\left[|f(\rho, \mathbf{r})|^2\right] d\rho d\mathbf{r} \tag{3.33}$$

As the field $f(\rho, \mathbf{r})$ is analytic, it can be expanded in an orthogonal set of functions

$$f(\rho, \mathbf{r}) = \sum_{i=0}^{\infty} f_i \phi_i(\rho, \mathbf{r}) \tag{3.34}$$

where the Fourier coefficients are determined by

$$f_i = \int_{(\rho, r)} f(\rho, \mathbf{r}) \phi^*(\rho, \mathbf{r}) d\rho d\mathbf{r} \tag{3.35}$$

and $\{\varphi(\rho, \mathbf{r})\}$ represents a complete set of orthonormal basis functions over $(\rho, \mathbf{r}) = V$. Substituting into the formula for m_v we can derive

$$m_v = \frac{\eta}{hf} \sum_{i=0}^{\infty} |f_i|^2 \tag{3.36}$$

We have transformed an integral of a random variable into an infinite sum of coefficients of a random variable. As such there is no guarantee that these coefficients are uncorrelated. However, if we have the covariance function of the random process, Eq. (2.34),

$$\Gamma_{r_0 r_0'}(\tau, \rho_0) = R_s(\tau) B_s(\rho_0) \tag{3.37}$$

then, using the Karhunen-Loeve expansion [1, p. 84]

$$\int_{(\rho, r)} \Gamma_{r_0, r_0'}(\mathbf{\rho_1}, \mathbf{\rho_2}; t_1, t_2) \phi_i(\mathbf{\rho_2}, t_2) d\mathbf{\rho_2} dt_2 = \lambda_i \phi_i(\mathbf{\rho_1}, t_1) \tag{3.38}$$

we can obtain an orthonormal set of functions $\{\phi(\rho, t)\}$ for which the coefficients are uncorrelated. If in addition the modulation process is Gaussian, then the coefficients are also independent.

3.5 Coherence separability

When we assume that the mutual coherence function is coherence separable, as we have, the Karhunen-Loeve equation also separates. This then yields two independent equations, one for time and one for space. Thus we have the two simultaneous equations

$$\int_\rho R_\tau(\boldsymbol{\rho_1},\boldsymbol{\rho_2})\varphi_i(\boldsymbol{\rho_2})d\boldsymbol{\rho_2} = \lambda_{\tau\,i}\varphi_i(\boldsymbol{\rho_1})$$

$$\int_r R_s(t_1,t_2)\phi_j(t_2)dt_2 = \lambda_{s\,j}\phi_j(t_1) \tag{3.39}$$

The field can now be expanded as

$$f(t,\mathbf{r}) = \sum_{i=1}^{\infty}\sum_{j=1}^{\infty} f_i\,g_j\phi_i(t)\varphi_j(\mathbf{r}) \tag{3.40}$$

With

$$f_i g_j = \int_{(\rho,r)} f(t,\mathbf{r})\phi_i{}^*(t)\varphi_j(\mathbf{r})dt d\mathbf{r} \tag{3.41}$$

The $\{\phi_i(\rho)\}$, $\{\varphi_j(\mathbf{r})\}$ are called eigenfunctions, and the $\{\lambda_i,\lambda_j\}$ $(\lambda_{ti} = |f_i|^2, \lambda_{sj} = |g_j|^2)$ are called eigenvalues. Thus for coherence-separable fields, the field expansion separates into temporal and spatial components. By substitution, and using Parceval's theorem, we have

$$m_v = \alpha\int_{A_d}\int_t^{t+T} I_d(\rho,\mathbf{r})d\rho d\mathbf{r} = \sum_i \lambda_{ti}\sum_j \lambda_{sj} \tag{3.42}$$

Example 3.1 We assume a rectangular receiving aperture ($A_r = 4ab$), produced by a coherence-separable field at a distance R. We assume that the source is described by an irradiance function $B(u)$, where $\mathbf{u}$ is a spatial vector in a plane parallel to the receiving aperture. The normalized spatial coherence function due to this source is given by the Fresnel-Kirckhoff approximation [5, p. 1; App. D, p. 421]

$$\tilde{R}(\mathbf{r_1},\mathbf{r_2}) = \left[\frac{\beta(\mathbf{r_1-r_2})}{\beta(0)}\right]\exp\left[j\frac{\pi}{R\lambda}\left(|r_1|^2-|r_2|^2\right)\right] \tag{3.43}$$

with

$$\beta(\mathbf{r}) = \int_u B(\mathbf{u})\exp\left[j\frac{2\pi}{R\lambda}\left(\mathbf{u}\cdot\mathbf{r}\right)\right]d\mathbf{u} \tag{3.44}$$

If the irradiance function subtends a small solid angle relative to the diffraction-limited field of view of the receiver, that implies coherence over the whole aperture with the source appearing as a point, and there is only one eigenfunction equal to $1/\sqrt{A_r}$, with eigenvalue λ_0. This is what happens in most cases. It is only when the transmission path disturbs the wavefront that we have deteriorated conditions. When the latter happens, the spatial coherence function has a correlation length that is less than the dimension of the aperture and we have more than one eigenvector, with corresponding eigenvalues. For the rectangular aperture, the eigenvectors can be approximated by two-dimensional plane waves,

$$\varphi_{q,l}(x,y) = \frac{\exp\left[\dfrac{j2\pi qx}{2a} + \dfrac{j2\pi ly}{2b}\right]}{\sqrt{A_r}} \tag{3.45}$$

where x and y are the components of r, and q and l are integers. This is commonly referred to as a plane wave decomposition, with the eigenfunctions sampling the irradiance function at a Nyquist rate. When the irradiance function subtends a large solid angle, the eigenvalues can be approximated by

$$\lambda(q,l) \approx \frac{\lambda^2 R^2}{\beta(0)} B\left(\frac{q\lambda R}{2a}, \frac{l\lambda R}{2b}\right) \tag{3.46}$$

Notice that the angular separation of the eigenvectors is $\frac{\lambda}{2a}$ in the x axis and $\frac{\lambda}{2b}$ in the y axis, with each spatial mode subtending an area equal to $\frac{\lambda R}{2a} \cdot \frac{\lambda R}{2b} = \frac{\lambda^2}{4ab} \cdot R^2$. Thus each of the modes appears to be a diffraction limited receiver in the direction of its eigenvector. This is shown in Figure 3.2.

If A_0 is the total area subtended by the receiver, then the number of modes (degrees of freedom) can be determined as

$$\text{Number of spatial modes} = \frac{A_0/R^2}{\lambda^2/4ab} + 1 = D_s \tag{3.47}$$

These are the spatial degrees of freedom in the system. From the perspective of background noise, the equipartition theorem implies that we get thermal noise in every mode. We saw in Chapter 1 that this noise was equal to

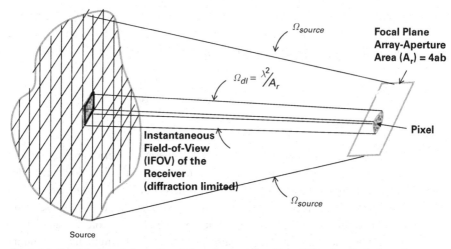

Figure 3.2 Graphical depiction of spatial modes contained in a diffraction limited receiver.

$$\frac{hf}{e^{hf/kT_0} - 1} \tag{3.48}$$

Hence the total amount of thermal noise received becomes

$$\frac{hf}{e^{hf/kT_0} - 1} \left[\frac{\Omega_{source}}{\lambda^2/A_r} + 1 \right] 2B_s \tag{3.49}$$

where we have included both spatial and temporal modes.

Example 3.2 Assume a circular aperture with radius r. We further assume that the source irradiance function $B(u)$ is homogeneous with circular symmetry. This allows us to compute

$$\beta(r) = \int_0^{r_0} B(u) J_0 \left(\frac{2\pi}{\lambda L} ur \right) u \, du \tag{3.50}$$

While there is no general solution to the Karhunen-Loeve equation, Slepian [6] has shown that for $B(u) = C_{constant}$, the eigenfunctions are generalized prolate spheroidal wavefunctions, and the eigenvalues depended upon the parameter

$$\Psi \equiv \frac{2\pi r r_0}{\lambda L} \tag{3.51}$$

He has also shown that for $\Psi \gg 1$, the number of significant eigenvalues is

$$D_s \approx \frac{\Psi^2}{4} = \frac{\pi^2 r_0^2 r^2}{(\lambda L)^2} = \frac{A_0 A_r}{\lambda^2 L^2} = \frac{A_0/L^2}{\lambda^2/A_r} \tag{3.52}$$

which yields the same result as the rectangular aperture. Here πr_0^2 is the area subtended by the object and πr^2 is the area of the aperture. In general, Helstrom [7] has suggested that for irregular apertures the dimension can be estimated as Eq. (3.52) with the area taken as

$$A_0 = \frac{\left[\int B(\mathbf{u}) d\mathbf{u} \right]^2}{\int B^2(\mathbf{u}) d\mathbf{u}} \tag{3.53}$$

What we see from these results is that as a coherent waveform passes through disruptive media, its plane of constant phase starts to distort. This distortion can be described by a coherence length, or a length at which portions of the wavefront start to decorrelate, and is inherent in the received field. For apertures whose dimension is less than a coherence length, the source will still appear to be a point with negligible degradation. However, for apertures whose dimension is greater than a coherence length, it will exhibit degradation in proportion to the ratio, in each dimension.

At the receiver, the plane wave first appears as a point source, and then starts to blur as it progresses. Finally, the blur fills the field of view of the (diffraction-limited) receiver, which is when the receiver dimension equals the coherence length. When it exceeds this point, more than one eigenfunction is needed to fully sample the field. Each of the Fourier coefficients will have independent statistics, so processing the outputs is a challenge. For the intensity-modulated communication system that we have discussed, combining is simple addition (power combining), or some other simple combination that weighs the strong coefficients more heavily. Power combining is also referred to as aperture averaging, and is only applicable to intensity-modulated beams. However, it also accumulates more background noise as shown in Eq. (3.49). This will be discussed further in later chapters. We point out here the work by Kennedy [8] on the theoretical bounds on both incoherent and coherent optical systems traversing random channels.

3.6   Summary

In this chapter we have taken the waveform generated in Chapters 1 and 2 through the receiving system into the photo-detector. We have shown the relationship between the field generated at the source, and the current generated in the detector. We have shown the existence of spatial degrees of freedom in a system, in addition to the normal temporal degrees of freedom. Specifically, for each temporal degree of freedom, there can be many spatial degrees. In later chapters we will address imaging and communications systems, and how one deals with this diversity.

References

1. R. M. Gagliardi and S. Karp. *Optical Communications*. Wiley Interscience, New York (1976).
2. S. Karp, R. M. Gagliardi , S. E. Moran and L. B. Stotts. *Optical Channels*. Plenum Press, New York (1988).
3. E. Whittaker. *A History of the Theories of Aether and Electricity*, Vols. I & II. Harper and Brothers, New York (1951).
4. A. Papoulis. *Probability. Random Variables, and Stochastic Processes*. McGraw-Hill (1965) pp. 300–304.
5. M. Born and E. Wolf. *Principles of Optics: Electromagnetic Theory of Propagation, Interference and Diffraction of Light*, 7th edn. Cambridge University Press, Cambridge (1999).
6. D. Slepian. Prolate spheroidal wave functions. *Bell System Technical Journal*, **43** (1964), p. 3009.
7. C. W. Helstrom. Modal decomposition of aperture fields in detection and estmation of incoherent objects. *Journal of the Optical Society of America*, **60**(4) (1970), p. 521.
8. R. S. Kennedy. Communication through optical scattering channels: an introduction. *Proceedings of the IEEE*, **58**(10) (1970), pp. 1651–1665.

4 Metrics for evaluating photo-detected radiation

In this chapter we will address the concepts of signal-to-noise ratio and contrast, and how they are used in electro-optic systems. We will cover analog modulation, digital modulation, background noise and a definition for contrast. In physics the distance to the receiver is usually designated by the symbol z, in communications by the range R. We will use both, but generally we will use z as the distance traversed in a bad channel and R as the distance traversed in a clear channel.

4.1   Reflective background noise

As we showed in Chapters 1 and 2, by radiometry and Maxwell's equations, the power collected by a receiver is proportional to the irradiance (intensity) and the area of the receiver. We also showed that if the source is, or appears, extended, we must also integrate the intensity over the area of the source that is viewed by the receiver. If the detector consists of a single element (pixel), this intensity would be the average radiated by the scene over the receiver field of view, which can be designated as $\bar{I}_s$, where

$$I_r = \frac{1}{R^2} \int I_s(\mathbf{r}) d\mathbf{r} \approx \bar{I}_s(r) \frac{A_d}{f_c^2}: \quad \text{extended:} \quad \left[= \frac{\bar{I}_s A_s}{R^2}: \quad \text{finite} \right] \tag{4.1}$$

and the power collected would be equal to

$$\bar{P}_r = \bar{I}_r A_r = \bar{I}_s(r) \frac{A_d A_r}{f_c^2}: \quad \text{extended:} \quad \left[= \frac{\bar{I}_s A_s A_r}{R^2}: \quad \text{finite} \right] \tag{4.2}$$

where A_d is the detector area, f_c is the focal length, R is the range to the target, and we have computed the result for finite and extended sources. Since $A_d R^2 / f_c^2$ is the area subtended by the target to the detector, this represents the area of integration. If on the other hand the detector consists of an array of $(M \times N)$ elements (pixels), then each pixel would sample a portion of the scene, described as $\bar{I}_s(x,y)$, or

$$\bar{I}_r(i,j) = \frac{1}{R^2} \int_{\frac{A_d R^2}{f_c^2}} I_s(\mathbf{r}_{i,j}) d\mathbf{r} \approx \bar{I}_s(x_i, y_j) \frac{A_d}{f_c^2} \tag{4.3}$$

with power collected equal to

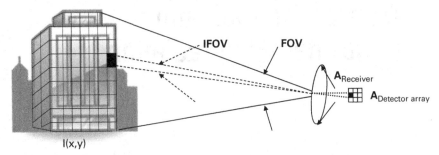

Figure 4.1 Field of view and instantaneous field of view of an image.

$$\bar{P}_r(i,j) = \bar{I}_r(i,j)A_r = \bar{I}_s(x_i,y_j)\frac{A_dA_r}{f_c^2} \tag{4.4}$$

and the array output would consist of

$$\sum_{i,j} \bar{P}_r(x_i,y_j) = \sum_{i,j} \bar{I}_r(x_i,y_j)A_r = \sum_{i,j} \bar{I}_s(x_i,y_j)\frac{A_dA_r}{f_c^2} \tag{4.5}$$

(Figure 4.1.) Summarizing, we have for the power received

$$\bar{P}_r(x_i,y_j) = \bar{I}_s(x_i,y_j)\frac{A_dR^2}{f_c^2}\frac{A_r}{R^2} = \bar{I}_s(x_i,y_j)\frac{A_dA_r}{f_c^2} \tag{4.6}$$

for the single pixel, and for the array

$$\sum_{i,j} \bar{P}_s(x_i,y_j) = \sum_{i,j} \bar{I}_s(x_i,y_j)\frac{A_dA_r}{f_c^2} \tag{4.7}$$

In the first case this would be the noise seen by a communications receiver looking at the transmitter with the scene as the background noise. In the second case this would be the image on the focal plane of a camera. Remember that the IFOV of the camera is determined by the size of the detector, A_d, and the focal length f_c. The lower bound on the IFOV is the defraction limit, or $A_d/f_c^2 \geq \lambda^2/A_r$, Eq. (2.60).

We have not discriminated between the reflection off a target and the thermal emission of a source. In the case of reflection $I_s(r_{i,j})$ would be determined by the intensity of the illuminator, I_o, multiplied by the reflection coefficient of the target $\rho_{i,j}$, or $I_s(r_{i,j}) = \rho_{i,j}I_o$. Notice that a scene is determined by the reflection coefficient of each of the elements in the scene. For targets that emit, an emmitance coefficient, $\varepsilon_{i,j}$, replaces $\rho_{i,j}$. In both cases the first Law of Thermodynamics requires that $\rho_{i,j} \leq 1$; $\varepsilon_{i,j} \leq 1$.

Example 4.1 Compute the received power by a pixel in a diffraction limited array, given that the spectral radiance is given by

$$N_e(r,\theta,\lambda) \tag{4.8}$$

Using the radiometry discussion in Chapters 1 and 2, we see that the integral over the bandwidth yields the radiance

$$N_e(r,\theta,\lambda)\Delta\lambda \tag{4.9}$$

We next compute the radiant intensity by integrating over the area of the extended source to obtain

$$\bar{N}_e\big((x_i,y_j),\theta,\lambda\big)\Delta\lambda\frac{\lambda^2 R^2}{A_r} \tag{4.10}$$

Finally we obtain the power by integrating over the solid angle projected from the receiver to the source

$$\bar{P}_r(x_i,y_j) = \bar{N}_e\big((x_i,y_j),\theta,\lambda\big)\Delta\lambda\frac{\lambda^2 R^2}{A_r}\left(\frac{A_r}{R^2}\right)$$

$$= \bar{N}_e\big((x_i,y_j),\theta,\lambda\big)\lambda^2\Delta\lambda \tag{4.11}$$

Comparing this to the earlier result (for a diffraction limited receiver) we have

$$\bar{P}_r(x_i,y_j) = \bar{I}_s(x_i,y_j)\frac{A_d A_r}{f_c^2} \Rightarrow \bar{I}_s(x_i,y_j)\lambda^2 \tag{4.12}$$

from which we can relate $\bar{I}_s(x_i,y_j)$ to $\bar{N}_e\big((x_i,y_j),\theta,\lambda\big)\Delta\lambda$.

4.2 Black-body (thermal) sources

If we assume that either the target or background comes from a black-body source, we can compute the power emitted, P_{BB}, from the source as the equipartition energy times the number of degrees of freedom. The quantum mechanical equipartition energy E_{eq} is (Eq. (1.17))

$$E_{eq} = \frac{hf}{e^{\frac{hf}{kT}}-1} \tag{4.13}$$

The number of temporal modes is equal to the bandwidth $2B$ per second. The number of spatial degrees of freedom in a pixel becomes

$$D = \frac{\text{Pixel} \quad \text{FOV}}{\text{Diffraction} \quad \text{limited} \quad \text{FOV}} = \frac{A_d f_c^2}{\lambda^2 A_r} = \frac{A_d A_r}{\lambda^2 f_c^2} \tag{4.14}$$

We can determine P_{BB} from a pixel to be (Eq. (3.49))

$$P_{BB} = \left(\frac{hf}{e^{\frac{hf}{kT}}-1}\right)\frac{A_d A_r 2B}{\lambda^2 f_c^2} \tag{4.15}$$

For thermal sources we can compute differential ΔP_{BB} due to a differential temperature ΔT, to be

$$\Delta P_{BB} = \frac{\partial P_{BB}}{\partial T}\Delta T = \left[\left(\frac{e^{\frac{hf}{kT}}\frac{(hf)^2}{kT^2}}{\left(e^{\frac{hf}{kT}}-1\right)^2}\right)\frac{A_r A_d 2B}{\lambda^2 f_c^2}\right]\Delta T \qquad (4.16)$$

Since we are using wavelength, $f = c/\lambda$ and $2B = \Delta f = c\Delta\lambda/\lambda^2$. Switching to wavelength we obtain for ΔP_{BB},

$$\Delta P_{BB} = \left[\left(\frac{e^{\frac{hc}{kT\lambda}}\frac{(hc)^2}{kT^2}}{\left(e^{\frac{hc}{kT\lambda}}-1\right)^2}\right)\frac{A_r A_d c}{\lambda^6 f_c^2}\right]\Delta\lambda\Delta T \qquad (4.17)$$

If we were looking to discriminate one element in the background from another, say P_1 and P_2, we can approximate this difference as $(P_1 - P_2) = \Delta P_{BB}$. This is accurate when looking for small differences in temperature. Similarly, for $(P_1 + P_2)$ we would use

$$P_1 + P_2 \cong P_{BB} = \left(\frac{hc^2}{e^{\frac{hc}{kT\lambda}}-1}\right)\frac{A_r A_d \Delta\lambda}{\lambda^5 f_c^2} \qquad (4.18)$$

Now thermodynamics enters in and says that this process, too, has an efficiency (relative to a pure black body) called emissivity. Hence the actual radiant intensity emitted is $\varepsilon(i,j)\,P_{BB}(i,j)$, and the variation or signal content of the scene is $\varepsilon(i,j)\Delta P_{BB}(i,j)$.

4.3   Mist, haze and fog

One might be tempted to say that either you can see the target in a fog, or you can't. We know, of course, that as you get closer to the target you can eventually see it. When the fog is thick enough so that the target can't be seen, the solution to the radiant transport equation yields a diffuse reflection coefficient that comes from the "infinite column". However, once you can see the target, the column is no longer infinite. What you then have is an attenuated beam reflecting off the target and getting further attenuated on the return, producing some amount of backscatter in the process. We will defer discussion of this situation for later in the book, but for our purpose here we will assume that the target can't be seen, and we have a diffuse return. We will assume that the return is homogeneous with a level equal to some incident spectral radiant value N_0 and diffuse reflection coefficient ρ_O.

$$P_O = \frac{\rho_O N_O A_r A_d \Delta\lambda}{f_c^2} \qquad (4.19)$$

for each pixel.

Notice that in these three sections range does not enter into the imaging system. This is because we did not keep resolution constant. Therefore as the range increases, the area covered by the pixel increases, and the number of pixels on target decreases as $1/R^2$.

4.4   Signal-to-noise ratio

In Chapter 3 we showed that the power spectral density of the photo-electron current, produced by the signal, was

$$S_x(\omega) \equiv \lim T \to \infty \frac{1}{2T} E\left[|X_T(\omega)|^2\right]$$

$$= e^2|H(\omega)|^2\left[\bar{n} + \bar{n}^2\left[2\pi\delta(\omega) + |S_m(\omega)|^2\right]\right] \tag{4.20}$$

where

$$\bar{n} = \lim T \to \infty \frac{\eta}{hfT} \int_{-t}^{t} I(t)A_r dt = \frac{\eta\bar{P}}{hf}$$

$$S_m(\omega) = \lim T \to \infty \left[\frac{1}{2T} E_m\left[|M_T(\omega)|^2\right]\right] \tag{4.21}$$

and

$$\frac{1}{2\pi}\int_{-\infty}^{\infty} S_m(\omega)d\omega = \lim T \to \infty \frac{1}{2T}\int_{-T}^{T} m^2(t)dt = \overline{m^2} : \quad m(t) \geq -1 \tag{4.22}$$

We also showed that in the absence of additive noise, the inherent "shot noise" created a discrete noise level equal to $2e^2\bar{n}B$, hence an inherent SNR equal to

$$SNR_{ql} = \frac{e^2\bar{n}^2\overline{m^2}}{2e^2\bar{n}B} = \frac{\eta\overline{m^2}\bar{P}_s}{2hfB} \tag{4.23}$$

We have identified $\bar{P}_s$ as the total signal power, and the signal-to-noise ratio as the "quantum-limited", or discrete, signal-to-noise ratio.

When we add background power to the signal and pass it through a photo-detector, in addition to the detected background power, there will be a mixing between the signal and the noise, which will increase the noise level [1, p. 148]. This mixing noise is a second-order effect and can be ignored. The additive noise will not be modulated and the contribution will be an increased shot noise level equal to $2e^2\bar{n}_bB$. Three sources of noise power have been identified above and can be selected to suit the circumstance.

In addition there exists internal dark current, which represents leakage from the detector, and kTB noise which comes from the thermal environment or the receiver after photodetection. Since we identify $i_{rms}^2 R = 2kTB$, we have $i_{rms}^2 = 4kTB/R$ and it can be eliminated with large resistors. Similarly, kTC noise, which is identified with switched capacitors, can be circumvented with double correlated sampling. This references the voltage output to the voltage after the creation of the pedestal. We will not include either in our computation of SNR.

4.5   Signal plus additive background noise

We have shown that the spectrum of the additive background noise is precisely the same as the signal, absent the modulation component $S_m(\omega)$. We next assume that the DC terms can be eliminated. Finally we assume the shot noise level $2e^2\bar{n}_bB$ for the background level, and $2e^2\bar{n}_d B$ for the dark current level. However, since the thermal noise is also Gaussian, we can safely conclude from section 3.2 that the total noise level appears Gaussian in nature. The signal to total background noise ratio becomes

$$SNR_b = \frac{e^2\bar{n}_s{}^2 m^2}{2e^2(\bar{n}_s + \bar{n}_b + \bar{n}_d)B} = \frac{\eta m^2 \bar{P}_s{}^2}{(\bar{P}_s + \bar{P}_b + \bar{P}_d)2hfB} \qquad (4.24)$$

(If the reader wishes to include the receiver thermal noise in computations, he would add $2kTB/R_\Omega = \bar{P}_T$ as a fourth current noise component in the denominator of Eq. (4.24).)

The noise equivalent power (NEP) of a detector is defined as the amount of signal power needed to produce an output SNR = 1, in a 1 Hz bandwidth, when the dark current noise is the dominant source of noise. (This is a measure usually applied to detectors before the advent of photo-electron counting receivers.) To determine this we have

$$1 = \frac{e^2\bar{n}_s{}^2 m^2}{2e^2(\bar{n}_d)B} = \frac{\eta m^2 \bar{P}_s{}^2}{2hf \bar{P}_d} \qquad (4.25)$$

From this we see that

$$\sqrt{\frac{2hf \bar{P}_d}{\eta m^2}} = \sqrt{\frac{2(hf)^2 i_{dc}}{\eta m^2 e}} = \bar{P}_s = NEP: \quad i_{dc} = \quad \text{dark current} \qquad (4.26)$$

Detectivity, D, is defined as $D = 1/NEP$, and D^* is the value of D for a 1 cm^2 detector.

As a final note we point out that some detectors such as photomultipliers and avalanche detectors have inherent gain. Thus when a photo-electron is released due to the field intensity, there are secondary interactions which cause a cascade of electrons to be released. This gain is not a constant G, but has some variance which is generally characterized as $G^{2+\delta}$, where $0 \le \delta \le 1$. This introduces the multiplier $G^{-\delta}$ into Eq. (4.24), while suppressing all other receiver circuit noise.

4.6   Signal-to-noise ratio for digital systems

Recall from Chapter 1 that Gaussian noise was the worst-case noise when computing capacity, and hence yielded a lower bound on capacity. In the case of the photo-detector, it is not necessary to assume Gaussian statistics. Devices exist that can yield close to perfect multiplication of single electron events. Therefore we are able to actually count the number of electrons that are released due to a signal impinging on the detector, hence measure the actual Poisson events in real time. The question that arises is "is there an optimum signal waveform that maximizes performance?" For Gaussian statistics the answer is no [2, p. 282]. For Poisson counting systems, the answer is yes.

When a discrete set of digital signals are transmitted, it is always necessary to establish a criterion for making a choice as to which was sent from the observed data. When the transmitted signals are equiprobable, the choice that maximizes the average probability of being correct is determined by the maximum likelihood function, Λ_i (or any monotonic function of it) [3, p. 214]. This is defined as

$$\Lambda_i \equiv p(k/i): \quad i = 1 \ldots M \qquad (4.27)$$

Binary signaling

Let us first consider the binary case in its simplest form, an on-off keyed system. Since the statistics are Poisson, we have for Λ_i,

$$P(k/i) = \frac{(\bar{n}_{s_i}T + \bar{n}_bT)^k}{k!} e^{-(\bar{n}_{s_i}T + \bar{n}_bT)} : \quad \bar{n}_{s_1} = \bar{n}, : \bar{n}_{s_2} = 0 \tag{4.28}$$

and

$$\ln \Lambda_i = k \ln \left(1 + \frac{k_{s_i}}{k_b} \right) - k_{s_i} - [k \ln (k_b) - \ln (k!) - k_b] \tag{4.29}$$

The term in brackets does not depend on i, therefore it will be the same for both signals and can be ignored. Hence the test becomes

$$\ln \Lambda_i = k \ln \left(1 + \frac{k_{s_i}}{k_b} \right) - k_{s_i} \tag{4.30}$$

which becomes

$$\ln \Lambda_1 > \ln \Lambda_2 \tag{4.31}$$

Or since $s_0 = 0$,

$$k \ln \left(1 + \frac{k_{s_i}}{k_b} \right) - k_{s_i} > 0 \Rightarrow k > \frac{k_{s_1}}{\ln \left(1 + \frac{k_{s_1}}{k_b} \right)} = k_T \tag{4.32}$$

So we measure the count k in the $(0,T)$ interval, and make the comparison above. If the test is positive we declare it signal one, if not signal two. The average probability of error, P_E, for the on/off waveform is [1, p. 218]

$$P_E = \frac{1}{2} \sum_{k=0}^{k_T} \gamma_{kk_T} \frac{(k_s + k_b)^k}{k!} e^{-(k_s+k_b)} + \frac{1}{2} \sum_{k=k_T}^{\infty} \gamma_{kk_T} \frac{(k_b)^k}{k!} e^{-(k_b)} : \gamma_{kk_T} = \frac{1}{2}; k = k_T$$

$$= 1; \text{otherwise} \tag{4.33}$$

which is plotted in Figure 4.2 as a function of the noise counts per bit.

Suppose now that we have the option of transmitting the same amount of energy in a smaller interval, Δt. This means that we need a higher peak power source, but both the energy and the average power remain the same. At the receiver we only look at the first Δt portion of the interval and ignore the remainder. Since the energy is the same, k_{s_1} also remains the same. On the other hand $k_{s_2} = \bar{n}_b \Delta t$, and the test becomes

$$k > \frac{k_{s_1}}{\ln \left(1 + \frac{k_{s_1}}{\bar{n}_b \Delta t} \right)} \tag{4.34}$$

In the limit as $\Delta t \to 0$, this test becomes $k > 0$. Thus as we make Δt small there is less and less chance for the noise to contribute, and we can use smaller values of k_{s_1}, hence less energy for the same outcome. This is shown in Figure 4.3. A vertical slice for P_E at a given value of k_{s_1} shows the value of reducing the noise count with low-duty-cycle operation.

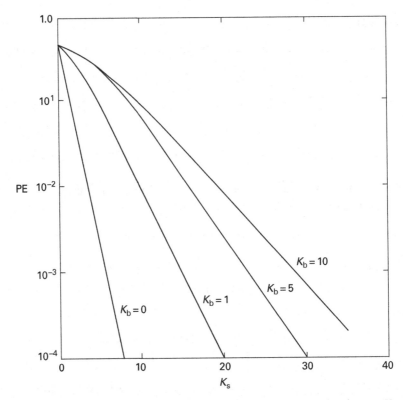

Figure 4.2 Bit error probability, OOK binary systems: K_s = average signal count, K_b = average noise count per bit interval [1].

Thus a low-duty-cycle signal is desirable in an environment characterized by Poisson noise. On the other hand, the on-off system is not desirable since the threshold can vary with changing channel characteristics and in general should be estimated before the comparison.

A modulation format that has a zero threshold is a Manchester code (2-PPM), where the slot interval T is divided into two halves, and the pulse is located at the leading edge of one of the two halves. Then the solution is to look in the leading edge (Δt) of the two halves and pick the one that has the largest count. The error probabilities for both cases can be readily derived [4, p. 208]. The probability of error for the 2-PPM system can be written in terms of the Marcum Q-function

$$Q(a,b) \equiv \int_b^\infty e^{-\left[\frac{(a^2 + x^2)}{2}\right]} I_0(ax)dx \qquad (4.35)$$

where $I_0(x)$ is the imaginary Bessel function. We then have

$$P_E = Q\left(\sqrt{2m_0}, \sqrt{2m_1}\right) - \frac{1}{2}e^{-(m_0 + m_1)} I_0\left(2\sqrt{m_0 m_1}\right) \qquad (4.36)$$
$$m_1 = K_s + K_b \quad : \quad m_0 = K_b$$

and is shown in Figure 4.4, for values of K_s, K_b.

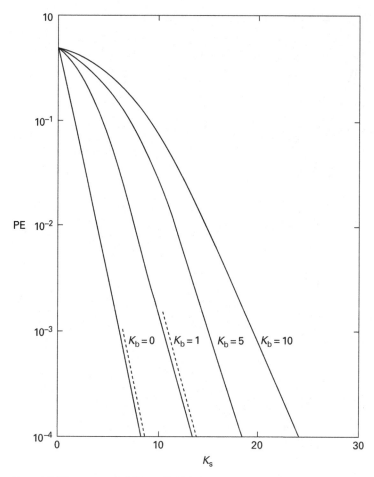

Figure 4.3 Bit error probability, PPM binary system: K_s = average signal count, K_b = average noise count per pulse interval [1].

M-ary signaling

Once we recognize that low duty operation is optimum, we are led to the fact that PPM also becomes the optimum M-ary waveform. In fact for the same time allocation T, we can locate the pulse in any one of $M = T/\Delta t$ slots providing $\log_2 M$ bits of information. The probability of making a word error, PWE, becomes [1, p. 263]

$$\text{PWE} = 1 - \frac{e^{-\left(k_s + M\bar{n}_b\Delta t\right)}}{M} - \sum_{k=1}^{\infty} \frac{(k_s + \bar{n}_b\Delta t)^k}{k!} e^{-\left(k_s + \bar{n}_b\Delta t\right)}$$

$$\left[\sum_{k=0}^{k-1} \frac{(\bar{n}_b\Delta t)^k}{k!} e^{-\left(\bar{n}_b\Delta t\right)}\right]^{M-1} \left(\frac{1}{Ma}\right)\left[(1+a)^M - 1\right] \qquad (4.37)$$

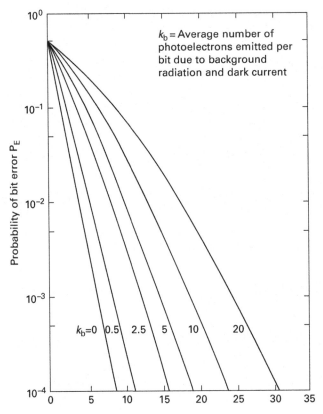

Figure 4.4 Average number of signal photoelectrons per bit k_s [4].

where

$$a \equiv \frac{(\bar{n}_b \Delta t)^k}{k! \displaystyle\sum_{i=0}^{k} \frac{(\bar{n}_b \Delta t)^i}{\Delta t!}} \tag{4.38}$$

An exemplary plot is shown in Figure 4.5. A more extensive set of curves can be found in [5, NASA]. However, there is a relationship between word error and bit error [6, p. 226)] which takes the form

$$P_E = \frac{1}{2}\left(\frac{M}{M-1}\right)\text{PWE} \tag{4.39}$$

To make a correct comparison, it should be done at the same data rate. First we recall (Eq. (1.25)) that the data rate can be written as

$$R = \frac{\log_2 M}{T} = \frac{\log_2 M}{M \Delta t} \text{ bits/second} \tag{4.40}$$

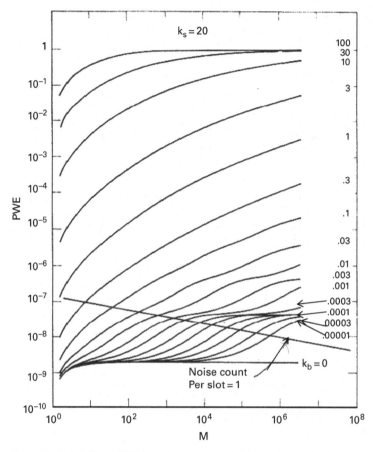

Figure 4.5 Probability of PPM word error versus M for various noise levels and signal count equal to 20 [1].

where

$$\Delta t = \frac{\log_2 M}{MR_0} \tag{4.41}$$

for any data rate R_0. Thus we see that

$$\bar{n}_b \Delta t = \left(\frac{\bar{n}_b}{R_0}\right)\left(\frac{\log_2 M}{M}\right) \tag{4.42}$$

as we let $\Delta t \to 0$. In Figure 4.5 we have also plotted the trajectory of PWE for $\bar{n}_b/R_0 = 1$ as we let $\Delta t \to 0$. This is summarized in Figure 4.6. We see that in this case for large M, there is about a 3 dB improvement in performance. Detailed analysis of M-ary systems can be found in [1, p. 257].

4.7   Signal-to-noise ratio of an image

In Chapter 2 we showed that for the Shannon definition of signal-to-noise ratio as used in the calculation for capacity, we had

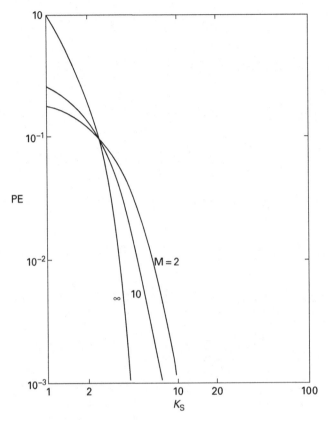

Figure 4.6 Equivalent bit error probability with block coded orthogonal signaling: K_s = signal count per word, $K_b = n_b$, $\emptyset_0 = 1$.

$$SNR = \frac{P}{N} = \frac{\Gamma_f(0)}{\Gamma_n(0)} = \frac{\int\limits_{-\infty}^{\infty} S_\tau(\omega)d\omega \int\limits_{-\infty}^{\infty} \tilde{I}_s(\mathbf{f})df}{\int\limits_{-\infty}^{\infty} S_n(\omega)d\omega \int\limits_{-\infty}^{\infty} \tilde{I}_n(\mathbf{f})df} \qquad (4.43)$$

Here we have separated the temporal variation of the signal as measured by the power spectral density of the intensity, multiplied by the signal-to-noise ratio in the angular distribution of the scene. Since in practice we will be looking at a single frame at a time, the temporal distribution is actually the energy in a frame, or power multiplied by frame time.

Consider the case where we scan across the image with a single pixel. This produces the upper trace in Figure 4.7. This would be an energy trace. Repeated scans of the same image would produce a power trace, but there would be no difference otherwise. If we remove the average value of this signal, the result would be the trace below it, where $m(x,y)$ is a signed number greater than or equal to minus one (≥ -1). $m(x,y)$ contains all the information about the content in the scene. We defined a positive average value for

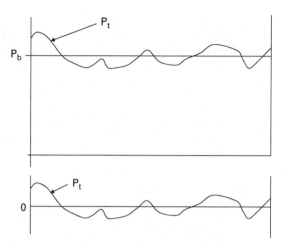

Figure 4.7 Intensity variation along a single row of pixels in an image.

this which was m. In the case of a communication signal, we called m^2 the modulation index. The same is true here, except that the information-bearing signal is an intensity, and hence we only need $m\bar{P}_s$ to show the effect on power. This merely reflects the AC component of the power which, we saw from Shannon, contains all the information in the signal, and is the image. The fact that we view this with a single scanned detector, a linear array, or a starring sensor does not alter the character of the signal. We assume the basic understanding is the same.

Thus in reconsidering the SNR_b when background only is present, we see that it factors into two components

$$SNR_b = \frac{\eta m^2 \bar{P}_s^{\,2}}{(\bar{P}_s + \bar{P}_b + \bar{P}_d)2hfB} = \left(\frac{m\bar{P}_s}{(\bar{P}_s + \bar{P}_b + \bar{P}_d)}\right)\left(\frac{\eta m\bar{P}_s}{2hfB}\right) \qquad (4.44)$$

The second term can be recognized as the quantum-limited signal-to-noise ratio. The first term we will define as contrast. When there is no additive noise, the contrast equals m, the spatial modulation index, and is the inherent contrast in the scene. When we have additive noise and dark current, we will write the contrast as

$$\left(\frac{m\bar{P}_s}{(\bar{P}_s + \bar{P}_b + \bar{P}_d)}\right) = \frac{m}{1 + \left(\frac{\bar{P}_b + \bar{P}_d}{\bar{P}_s}\right)} = \frac{\text{Inherent}\quad\text{Contrast}}{\text{Contrast}\quad\text{Attenuation}} \qquad (4.45)$$

We will make a few points. There are other definitions of contrast, but they were all defined when the medium for imaging was film. It was difficult to understand that there was a "shot noise" riding on the AC component. In fact the noise introduced by the graininess of the film masked this out. But, as in an amplitude-modulated signal, it is the signal-to-noise ratio that determines information content, not just the amplitude of the return. With electronic media for storage, the DC term can be eliminated (with some restoration for viewing), and the fundamental properties of the image observed. With this definition we can link contrast with SNR and gain greater understanding of what determines true visibility. To distinguish between the modulation indices for temporal and spatial processes, we will continue to define the temporal modulation index as m, and the spatial modulation index, the average inherent contrast, as C_0.

The denominator in the term for contrast merely contains the sum of the variances of the contributing sources. The signal contribution to power comes from a time (space) varying mean not a variance, in the derivation of the power spectral density. To see this notice that if we were to repeatedly scan this same line with a single detector, we would continually get the identical result. However, the shot noise generated by the flow of electrons would have a different state with each scan. Thus one could run a test to measure the signal-to-noise ratio as a function of the number of scans N, and it would increase as $10\log_{10}N$ dB (assuming perfect calibration, hence no pattern noise).

Example 4.2 What would be the contrast of a periodic modulation that was on a fraction α of the period with value P and off the remaining time with zero value? If we denote the periodic wave by $P(x)$, then the average value would be αP. The function $m(x)$ would be $(1 - \alpha)P$ for the fraction α, and $(-\alpha P)$ for the fraction $(1 - \alpha)$.

$$\int_0^1 m^2(x)dx = (1-\alpha)^2\alpha + \alpha^2(1-\alpha) = (1-\alpha)\alpha \tag{4.46}$$

The average inherent contrast C_0 becomes $\sqrt{(1-\alpha)\alpha}$, which is maximum for $\alpha = 1/2$, for which the average inherent contrast $C_0 = 1/2$. Thus scenes that have a large percentage of bright areas $\alpha \to 1$ have low inherent contrast, and scenes that have a low percentage of bright areas $\alpha \to 0$ have low inherent contrast. This is defined across the whole scene. Within the scene there can be areas of high inherent contrast that can be extracted, and is why local area processing (LAP) is a valuable tool in image processing.

Example 4.3 Compute the SNR and the inherent contrast of a thermal scene.
We will identify $C_0\bar{P}_s$ with ΔP_{BB}, Eq. (4.17), or

$$C_0\bar{P}_s = \Delta P_{BB} = \left[\left(\frac{e^{\frac{hc}{kT\lambda}}\frac{(hc)^2}{kT^2}}{\left(e^{\frac{hc}{kT\lambda}}-1\right)^2}\right)\frac{A_rA_dc}{\lambda^6f_c{}^2}\right]\Delta\lambda\Delta T \tag{4.47}$$

Similarly, we will identify $\bar{P}_b$ with P_{BB}, (4.15), or

$$\bar{P}_b = P_{BB} = \left(\frac{hc}{e^{\frac{hf}{kT}}-1}\right)\frac{A_d A_r c \Delta \lambda}{\lambda^5 f_c^2} \tag{4.48}$$

In both terms we have kept the bandwidth in wavelength. The inherent contrast in a thermal image would then be the ratio, or

$$C_{Thermal} = \frac{\left[\left(\frac{e^{\frac{hc}{kT\lambda}}\frac{(hc)^2}{kT^2}}{\left(e^{\frac{hc}{kT\lambda}}-1\right)^2}\right)\frac{A_r A_d c}{\lambda^6 f_c^2}\Delta\lambda\Delta T\right]}{\left(\frac{hc}{e^{\frac{hf}{kT}}-1}\right)\frac{A_d A_r c \Delta\lambda}{\lambda^5 f_c^2}} = \left[\left(\frac{e^{\frac{hc}{kT\lambda}}\frac{(hc)}{kT^2}}{\left(e^{\frac{hc}{kT\lambda}}-1\right)}\right)\frac{\Delta T}{\lambda}\right] \tag{4.49}$$

The signal-to-noise ratio then becomes

$$SNR_{Thermal} = \left[\left(\frac{e^{\frac{hc}{kT\lambda}}\frac{(hc)}{kT^2}}{\left(e^{\frac{hc}{kT\lambda}}-1\right)}\right)\frac{\Delta T}{\lambda}\right]\frac{\eta\Delta P_{BB}\lambda}{hc2B_e} = \left[\left(\frac{e^{\frac{hc}{kT\lambda}}}{\left(e^{\frac{hc}{kT\lambda}}-1\right)}\right)\frac{\Delta T}{kT^2}\right]\frac{\eta\Delta P_{BB}}{2B_e} \tag{4.50}$$

Finally we have

$$SNR_{Thermal} = \eta\left[\left(\frac{\left(e^{\frac{hc}{kT\lambda}}\frac{(hc)}{kT^2}\right)^2}{\left(e^{\frac{hc}{kT\lambda}}-1\right)^3}\right)\frac{\Delta T^2 A_d A_r \Delta\lambda\tau_e}{\lambda^7 f_c^2}\right] \tag{4.51}$$

where $\frac{1}{2B_e} = \tau_e$. For diffraction limited operation we have

$$SNR_{Thermal-DL} = \eta\left[\left(\frac{\left(e^{\frac{hc}{kT\lambda}}\frac{(hc)}{kT^2}\right)^2}{\left(e^{\frac{hc}{kT\lambda}}-1\right)^3}\right)\frac{\Delta T^2 \Delta\lambda\tau_e}{\lambda^5}\right] \tag{4.52}$$

Example 4.4 How many bits are required in an analog-to-digital converter (ADC) to be able to view a scene at a defined level of contrast?

Contrast has been defined as

$$\text{Contrast} = \frac{C_0\bar{P}_s}{\bar{P}_s + \bar{P}_b + \bar{P}_d} \tag{4.53}$$

where the powers are the input variables to the detector. As shown in Chapter 3, the photo-detector is a power (energy) detector that is intensity in, electrons out. If we have an n-bit ADC on the current output of the photo-detector, we would normally assume $20\log(2^n) \cong 6n$ dB or 6 dB per bit. This specifies the quality of the output signal.

However, since we are not interested here in the properties of the output current, but rather the properties of the input contrast, we obtain this by computing $10 \log (2^n) \cong 3n$ dB or only 3 dB per bit. Thus, for example, if we define visibility as 2% contrast or -17 dB, this corresponds to 5.66 bits. To resolve variations below 2% requires an ADC with greater than six bits of resolution. If we want to resolve images with three bits of resolving power at 10 dB below visibility, we require $5.66 + 3 + 10/3 = 12$ effective bits of resolution. Thus we compute the number of bits of resolution, ρ, needed as

$$\rho = 3 - \frac{10 \log_{10}(\text{Contrast})}{3} \tag{4.54}$$

Since the contrast varies with the propagation distance R, so does the number of bits required. Since it is necessary to simultaneously maintain a suitable SNR, there will always be a limit to the range, and hence the contrast enhancement achievable. This is displayed graphically in Figure 4.8.

Formally, if we take $10 \log$ (SNR) in Eq. (4.44), we get

$$10 \log (SNR_b) = 10 \log \frac{\eta C_0{}^2 \bar{P}_S{}^2}{(\bar{P}_S + \bar{P}_b + \bar{P}_d) 2 h f B}$$

$$= 10 \log \left(\frac{C_0 \bar{P}_S}{\bar{P}_S + \bar{P}_b + \bar{P}_d} \right) + 10 \log \left(\frac{\eta C_0 \bar{P}_S}{2 h f B} \right) \tag{4.55}$$

and arrive at the same result for contrast. These results will be discussed further in Chapter 5.

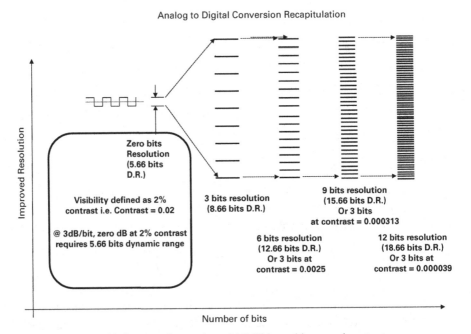

Figure 4.8 Relationship between the number of ADC bits and improved contrast.

4.8 Summary

In this chapter we have introduced the concepts necessary for electro-optic system design. These include the computation of signal power, background power, signal-to-noise ratio and contrast for both irradiated and thermal systems. We have covered both analog and digital systems. Finally, we have shown what the requirements are for an ADC to obtain a defined level of contrast in an image. In addition to this level of contrast, we have quantified that there must simultaneously be sufficient SNR for proper viewing.

References

1. R. M. Gagliardi and S. Karp. *Optical Communications*. Wiley Interscience. New York (1976).
2. R. M. Gagliardi. Introduction to Communications Engineering. 2nd edn., John Wiley & Sons, New York (1998).
3. J. M. Wozencraft and I. M. Jacobs. *Principles of Communication Engineering*. John Wiley & Sons, New York (1967).
4. W. K. Pratt. *Laser Communication Systems*. John Wiley & Sons, New York (1969).
5. S. Karp, M. G. Hurwitz and R. M. Gagliardi. Error Probabilities for Maximum Likelihood Detection of M-Ary Poisson processes in Poisson noise. NASA, Tech. Note TN-D-4721 (October 1968).
6. A. J. Viterbi. *Principles of Coherent Communications*. McGraw-Hill, New York (1966).

5 Contrast, visibility and imaging

5.1 Background

The study of contrast and visibility for imagery is not new [1]. In fact it has been a scientific topic for over 2000 years, and originated when the Greek astronomer Hipparcus first classified stars by their intensity in 120 BC. He ranked them on a scale from 1 to 6, with 1 being the brightest he could see, and 6 the dimmest. There were no improvements made for the next three centuries, when Ptolomy classified 1022 stars in his classical work, the "Almagest". With the development of the telescope by Galileo, there was the discovery of a vast number of stars not visible with the naked eye. The first formal classification was made by Pogdon [2], who classified them by magnitude. He had the same six classifications, with 1 being the brightest and 6 being the faintest one visible with the naked eye. He set a scale of 100 with five steps in between at a factor $\sqrt[5]{100} = 2.512$ steps apart, which continues today for dimmer stars (> 6: sub-visible). One of the first metrics describing the lowest level of visibility was called the contrast threshold ε, or its inverse, contrast sensitivity. Generally speaking, this is the level at which the image in a scene becomes indiscernible from the background. Many measurements have been made of the contrast threshold, with a very extensive set made by Blackwell [3] as far back as 1946. In addition, there have been many definitions of contrast, the two most used being:

Michelson formula:

$$\text{Contrast} = \frac{P_{\max} - P_{\min}}{P_{\max} + P_{\min}} \tag{5.1}$$

(this is the definition we have used), and the Weber definition of contrast:

$$\text{Contrast} = \frac{P_{\max} - P_{\min}}{P_{\max}} \tag{5.2}$$

Technically, it is a measurement of the instantaneous dynamic range that can be resolved within a scene by the human eye. However, the absolute dynamic range of the eye is enormous, well over six orders of magnitude, whereas the range of contrast sensitivity is about 17 dB. Thus the capability of the human eye is akin to having an excellent AGC. It has been shown that contrast efficiency varies among individuals. Therefore, its understanding has gained importance as a measure of the visual capability of humans in areas where sight is required, such as driving. It is also important for aviation in determining safe landing conditions, where the concept of meteorological visibility has been developed. For this latter application, the most accepted value for zero contrast appears to be 0.02, although it can vary from 0.018 to 0.03. The "meteorological visibility" in distance is defined as x_o, where it satisfies the relationship

$$e^{-cx_o} = 0.02 \tag{5.3}$$

or $x_o = 3.912/c$, where c is the extinction coefficient of the environment. The smaller the extinction coefficient, the greater the visibility is. The contrast definition used here was the Weber definition.

For the former application, many experiments have been made to try to measure the contrast ratio, but it not only varies among individuals, so too does the metric. The Michelson definition appears to be correct with normal individuals, whereas the Weber definition appears best for individuals with scattering within the eye due to cataracts, etc. [4].

As we described earlier, normal propagation follows the range equation when there is nothing in the path of the propagating wave. However, when there are particulates such as molecules, aerosols, dust, etc. in the atmosphere these create deleterious effects in the form of absorption and scattering which impede normal propagation. The first step in classifying propagation effects starts with identifying inherent properties, which started with Mie [5].

5.2 Mie scattering

To understand how losses occur in the atmosphere, water or any other medium dominated by the presence of particulates it is important to have a working knowledge of Mie theory. A full discussion of Mie theory is beyond the scope of this book. Good discussions of this topic can be found in [6–11]. Briefly stated, Mie theory is the solution to Maxwell's equations for a plane wave impinging upon a homogeneous sphere having a complex refractive index, $n = n_r - jn_i$, a diameter D, by a waveform with a wavelength λ. If we define $S_1(n, x, \theta)$ as the perpendicular component of the scattered field, and $S_2(n, x, \theta)$ the parallel component, they can be written as

$$S_1(n, x, \theta) = \sum_{k=1}^{\infty} \frac{2k+1}{k(k+1)} [a_k(n, x)\pi_k(\theta) + b_k(n, x)\xi_k(\theta)] \tag{5.4}$$

$$S_2(n, x, \theta) = \sum_{k=1}^{\infty} \frac{2k+1}{k(k+1)} [a_k(n, x)\xi_k(\theta) + b_k(n, x)\pi_k(\theta)] \tag{5.5}$$

The $a_k(n, x)$ and $b_k(n, x)$ are called the Mie coefficients, and the $\pi_k(\theta)$ and $\xi_k(\theta)$ are spherical harmonics which can be written as Legendre polynomials. The Mie parameter x is defined as

$$x = \frac{\pi D}{\lambda} \tag{5.6}$$

For mono-dispersed particles, all spherical with fixed diameter, Mie theory gives an exact solution. For our purposes it is sufficient to know that this solution consists of a scattering cross section σ_s, an absorption cross section σ_a, and an extinction cross section $\sigma_e = \sigma_s + \sigma_a$, for each particulate. It also produces a normalized scattering phase function, $p(\theta)$, which can be considered to be the probability density of the scattering occurring in any direction covering 4π. The extinction cross section can be written as

$$\sigma_e(m, x) = \frac{2\pi}{k_0^2} \sum_{i=1}^{\infty} (2i + 1)\text{Re}\{a_i + b_i\}$$

$$= \frac{\pi D^2}{4} Q_e(m, x) \tag{5.7}$$

with $Q_e(m, x)$ being the extinction efficiency, and $k_0 = 2\pi/\lambda$. Similarly, the scattering cross section becomes

$$\sigma_s(m, x) = \frac{2\pi}{k_0^2} \sum_{i=1}^{\infty} (2i + 1)\left\{|a_i|^2 + |b_i|^2\right\}$$

$$= \frac{\pi D^2}{4} Q_s(m, x) \tag{5.8}$$

with the absorption cross section $\sigma_a(m, x) = \sigma_e(m, x) - \sigma_s(m, x)$. The ratio $\omega_0 = Q_s/Q_e$ is defined as the single scatter albedo. The normalized scattering phase function becomes

$$p(\theta) = \frac{2\pi|S_1(\theta)|^2}{x^2 Q_s} + \frac{2\pi|S_2(\theta)|^2}{x^2 Q_s} \tag{5.9}$$

with

$$\int_\Omega \frac{p(\theta)}{4\pi} d\Omega = 1 \tag{5.10}$$

In Figure 5.1, we show two representative solutions of these Mie efficiencies, one having a complex index close to zero (no absorption) (5.1a) and one having a finite complex value (5.1b). The first one, having no loss, exhibits pure scattering and is representative of scattering through clouds at visible wavelengths, where water deposits on aerosols, creating near spherical droplets. Nevertheless, there are similarities between the two curves. At short wavelengths both exhibit Rayleigh scattering [12] $\sim 1/\lambda^4$, and tend to scatter omni-directionally. This is more representative of molecular scattering, but not aerosols, which are appreciably larger. At the shorter wavelengths both curves approach their physical area and scatter in the forward direction. (There is a factor of two difference between the actual cross section and the asymptotic value, which is attributed to diffraction [8].)

To go the next step it is necessary to know the density of particles and a size distribution for the particles. Having such information allows the formulation of volume scattering coefficients, which are averages over particle density and particle size distributions [9]. These quantities are volume absorption a, volume scattering b, and volume extinction c, all having dimensions of per unit length, with $a + b = c$. Thus a ray of light entering a medium will be extinguished by e^{-cz} over the path length z, e^{-az} by absorption, and e^{-sz} due to scattering. The absorption is non-recoverable, but the scattering is merely

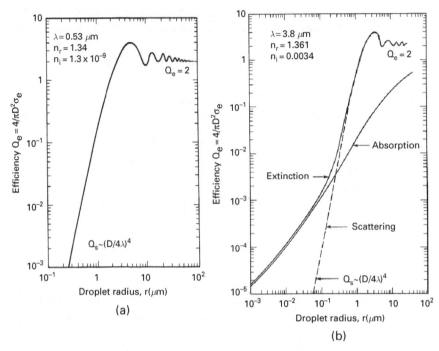

Figure 5.1 Absorption, scattering, and extinction coefficients versus particle size for 0.53 μm and 3.8 μm wavelength light.

redirected and under certain conditions may be recovered. In general for poly-dispersed distributions we have

$$P_1(\theta) = \frac{4\pi^2}{k_0^2 b} \int_0^\infty \rho(n, x) |S_1(n, x, \theta)|^2 dx$$

$$P_2(\theta) = \frac{4\pi^2}{k_0^2 b} \int_0^\infty \rho(n, x) |S_2(n, x, \theta)|^2 dx$$

$$b = \frac{\pi}{k_0^2} \int_0^\infty x^2 \rho(n, x) Q_s(n, x) dx$$

$$c = \frac{\pi}{k_0^2} \int_0^\infty x^2 \rho(n, x) Q_e(n, x) dx$$

$$a = \frac{\pi}{k_0^2} \int_0^\infty x^2 \rho(n, x) Q_a(n, x) dx \qquad (5.11)$$

The total concentration of particles is given by

$$N_p = \frac{1}{k_0} \int_0^\infty \rho(x) dx \qquad (5.12)$$

Instruments have been constructed to measure these quantities *in situ*, with limited success. The first such instrument was the Secchi disc used to measure extinction in water. This was a white disc that was lowered over the side of a ship to a depth at which it disappeared, the Secchi depth. Since this was the depth at which contrast was lost, it was actually a depth, D_s measured in meters, where $e^{-cD_s} = 0.02$. This occurred when $c = 3.912/D_s$. Actual attenuation (extinction) meters were next developed where a narrow beam was transmitted over a calibrated distance and the receiver had a very narrow field of view, hence receiving no scattered radiation and making a direct measurement of extinction. More recently, absorption meters have been developed which can measure a directly, hence b can be determined as $c - a$ [13]. In inhomogeneous water these parameters are depth-dependent.

In the atmosphere measurements are more difficult to make and the determination of a, b and c is more often done with models of particle density and size distributions [9]. Then by use of Monte Carlo simulations, corroboration with measured data can be made [14]. Another important application is similar to the attenuation meter. When actively interrogating the atmosphere with a collimated pulsed source, the returned beam attenuates as e^{-2cz}. Looking at the backscattered return with a narrow field of view allows a measurement of c. By exploiting certain physical and chemical characteristics of the backscatter, meteorological information can be obtained from these data, and the chemical's distribution in the atmosphere can be determined. Also, constituents of the atmosphere can be observed as a function of distance. This will be discussed later in Chapter 9 on lidar.

5.3 Radiative transport and imaging

In Chapter 4 we developed formulas for both signal-to-noise ratio and contrast and the relationship between the two. These expressions were developed for a benign environment, where the target was not obscured by obstacles in the transmission path. While these expressions are correct, they are incomplete. In this chapter we will draw upon the wealth of work created on radiative transport through a multiple scattering medium, and enrich these expressions accordingly. The original work in this area was started by Siebert Q. Duntley [15–17] at the Scripps Oceanographic Institute, where he investigated optical propagation in the atmosphere and in the ocean. This work is summarized in *Marine Optics* [18]. Other summaries of this work can be found in Karp, S., *et al.* [19]. In this chapter we will only skim these results as they apply to our goal of presenting a clearer description of contrast, and how modern signal processing can be used to apply this work to optical communications, lidar and low-visibility imaging.

We must first identify what happens to the image. It has been shown by Lutomirski and Yura [20] that the mutual coherence function, $M(\rho,\mathbf{r})$, for the forward scattering approximation of the radiative transport equation becomes

$$M(\rho,z) = e^{-az}e^{[-bz(1-\beta(\rho))]}$$

$$\beta(\rho) = \int_0^\pi \left[\int_0^1 J_0\left(\frac{2\pi\rho\tan(\theta)u}{\lambda}\right) du \right] p(\theta) \sin\theta d\theta \quad (5.13)$$

Using this result, one can show that the asymptotic value as $\rho \to \infty$ is

$$M(\rho,z) = e^{-az-\left(\rho/r_0\right)^2}; \quad \rho << \frac{r_0}{\sqrt{bz}}$$

$$= e^{-cz} \quad : \quad \rho << \frac{r_0}{\sqrt{bz}} \Rightarrow z >> \frac{\left(r_0/\rho\right)^2}{b} \tag{5.14}$$

where

$$r_0 = \lambda \Big/ \sqrt{bz\theta_{rms}}$$

$$\theta_{rms}^2 = 2\pi \int_0^\pi \theta^2 p(\theta)\sin\theta d\theta \tag{5.15}$$

e^{-cz} represents the loss in the unscattered, or spatially coherent, portion of the beam. This verifies that there is a component of the unscattered image which is preserved. Thus a point source will retain its integrity over the path length z. Furthermore an array of points, an image, will also retain its integrity over this path, although it will be swamped with scattering. What we showed in Chapter 4 was how to correctly define the signal-to-noise ratio of the image and relate it to the contrast of the scene. The ability to image through a scattering medium is directly proportional to one's ability to isolate the energy associated with the image and only include the noise associated with this isolation. Since the image is progressing through the medium as though there were no scattering, $b=0$, it must also be spreading geometrically and suffer a $1/R^2$ loss. (We distinguish between the total path from transmitter to receiver, R, and the extent through which the field encounters the scattering environment, z.) This can be visualized as going through a thick pane of colored glass that only absorbed. Actually, for this case it can be shown that that the asymptotic value as $\rho \to 0$ is

$$M(\rho,z) \Rightarrow e^{-az} \tag{5.16}$$

which is equivalent to $b = 0$.

A similar result to Eq. (5.16) had been obtained in the earlier work in radiative transport [15,16,18], where it was shown that the ratio of apparent contrast to inherent contrast decreased by the factor e^{-cz}, or

$$\frac{C_A(z,\theta,\phi)}{C_0(0,\theta,\phi)} = e^{-cz} \tag{5.17}$$

Here the Weber definition of contrast was used. What was also introduced in this work was the concept of apparent properties. The inherent property of contrast at the source is diminished due to contrast attenuation. Hence the inherent radiance is reduced to the apparent radiance. Another important apparent property that was introduced was the diffuse attenuation coefficient, k. The meaning of k is the steady-state loss in radiance, per unit length, in an infinite homogeneous medium. This includes all of the absorption loss plus some of the scattered radiation that goes in a different direction. As such, $a \leq k \leq c$. k would be used in calculations such as when the Sun's rays impinge on the top of the atmosphere and are observed on the Earth's surface, or when the Sun's rays impinge on the surface of the ocean and are observed from below the surface. Also a diffuse reflection

Figure 5.2 Typical imaging geometry in a scattering environment.

coefficient was introduced. Thus for example, suppose an object was completely immersed in a fog and the Sun illuminated both the object and the fog with the object not visible from the receiver (Figure 5.2). The Sun would still illuminate the fog and the receiver would get the backscatter from the fog, yielding a background radiance measuring the return from an effectively infinite column. The reflection coefficient of this return is called the diffuse reflection coefficient.

Example 5.1 What is the effect of the scatter channel in Eq. (5.13) on the propagation of an image for large z? From Eq. (2.54) we see that the spatial coherence function blurs the image by decreasing the coherence. However, the asymptotic limit for large z is a constant e^{-cz}. This implies that there is a latent image buried in the scattered radiation waiting to be found, but diminished by extinction.

5.4 Information content of images

In Chapter 3 we showed how an intensity that varied with time and space was transformed into an electronic process by means of a quantum detector. We then showed that we could follow the electromagnetic field through this detection process and reconstitute the properties in both time and space. We invoked coherence separability, which allowed us to treat the time and space properties independently. This is valid for static and slowly varying scenes, and is why image compression techniques succeed. This also follows Shannon's argument of $2BT$ dimensions in his original derivation of information capacity. In addition we showed that the inclusion of an image adds additional dimension to the information content of the scene and using time–space separability we can identify the mathematical structure of the image, and hence its information content. Our derivation of the detection process was performed during a time τ_e and over an area A. This allows us to reformulate the previous derivation during a period of time and over a finite area.

Recall that the probability density of the Poisson process generated (probability of k electrons) was (Eq. (3.2))

$$P_k(k) = \frac{(m_v)^k}{k!} e^{-m_v}$$

where the variable m_v satisfied Eq. (3.3),

$$m_v = E_k[k] = \frac{\eta}{hf} \int_{A_d} \int_t^{t+\tau_e} I_d(\rho, \mathbf{r}) d\rho d\mathbf{r} = \int_{A_d} \int_t^{t+\tau_e} n(\rho, \mathbf{r}) d\rho d\mathbf{r}$$

We then associated a shot noise process[1] $x(\mathbf{r},t)$ (the current produced by an ensemble of electrons) with the Poisson process (Eq. (3.12))

$$x(\mathbf{r}, t) = \sum_{j=1}^{k} h[(\mathbf{r} - \mathbf{r}_j); (t - t_j)] \quad : \quad \mathbf{r}_j, t_j \in [A; (t, t + \tau_e)]$$

where each function $h(\mathbf{r},t)$ represented the effect of an electron located in the space AT, satisfying Eq. (3.11),

$$\iint h(t, \mathbf{r}) dt d\mathbf{r} = e$$

and being identically distributed (Appendix A) with probability

$$p_V(\mathbf{r}_j, t_j) = \frac{n(\mathbf{r}_j, t_j)}{\int_A \int_t^{t+T} n(\mathbf{r}, t) d\mathbf{r} dt} \tag{5.18}$$

Now if we assume an interval τ_e small enough so that the scene is static at time $\tilde{t}$ (mean value theorem for integrals), we have

$$p_V(\mathbf{r}_j, \tilde{t}) \simeq \frac{n(\mathbf{r}_j, \tilde{t})}{\tau_e \int n(\mathbf{r}, \tilde{t}) d\mathbf{r}} = \frac{n(\mathbf{r}_j, \tilde{t})}{E[k]|_{\tilde{t}, \tau_e, A}} \equiv p_r(\mathbf{r}_j) \tag{5.19}$$

---

[1] Here we have assumed that the shot noise process is determined by the flow of a finite stream of electrons. If the radiation field is modeled as a stream of classical wave packets described as electromagnetic fields, this would introduce an additional shot noise process into the current which is separate, distinguishable, and measureable, e.g., [33,34].

with the shot noise process taking the form

$$x(\mathbf{r}, \tilde{t}) = \sum_{j=1}^{k} h[(\mathbf{r} - \mathbf{r}_j); \tilde{t}] \quad : \quad \mathbf{r}_j \in [A] \tag{5.20}$$

The key point is to now recognize that since $n(\mathbf{r}_j, \tilde{t})$ is an intensity, $n(\mathbf{r}_j, \tilde{t}_j) \geq 0$, $p_V(\mathbf{r}_j, \tilde{t})$ can be rewritten (as before in Chapter 3)

$$p_V(\mathbf{r}_j, \tilde{t}) \equiv \frac{\bar{k}_{\tilde{t}\tau_e A}[1 + m(\mathbf{r}_j, \tilde{t})]}{\bar{k}_{\tilde{t}\tau_e A} A} = \frac{[1 + m(\mathbf{r}_j, \tilde{t})]}{A} \tag{5.21}$$

with the variation $m(r, \tilde{t})$ representing the image in the focal plane (Figure 4.7), $\bar{k}_{\tilde{t}AT}$ the average number of electrons at the time $\tilde{t}$, over the area A, and time interval τ_e with

$$\int m(\mathbf{r}, \tilde{t}) d\mathbf{r} \equiv 0 \tag{5.22}$$

The Fourier transform of the process $x(r, \tilde{t})$ becomes

$$X_{\tilde{t}}(\mathbf{f}) = H_{\tilde{t}}(\mathbf{f}) \sum_j e^{j\mathbf{f} \cdot \mathbf{r}_j} \tag{5.23}$$

and the expected energy spectrum for the process becomes

$$E[|X_{\tilde{t}}(\mathbf{f})|^2] = |H_{\tilde{t}}(\mathbf{f})|^2 [\bar{k}_{\tilde{t}} + \bar{k}_{\tilde{t}}^2 [2\pi\delta(\mathbf{f}) + \Phi_m(\mathbf{f}, \tilde{t})] \tag{5.24}$$

where we have assumed the normalization

$$\int h(t, \mathbf{r}) dt = e$$

so that

$$\int h(t, \mathbf{r}) d\mathbf{r} = h(t) H_{\tilde{t}}(0) \Rightarrow H_{\tilde{t}}(0) = 1 \tag{5.25}$$

Notice that, as before, there are three terms in this spectrum all multiplied by the spectrum of $h(\mathbf{r})$, with $\bar{k}_{\tilde{t}}$ the average count at the time $\tilde{t}$, over the area A, during the time interval τ_e. This is then the transform of the scene in the focal plane. By Parseval's theorem,

$$\frac{1}{2\pi} \int_{-\infty}^{\infty} \Phi_m(\mathbf{f}, \tilde{t}) d\mathbf{f} = \frac{1}{A} \int_A E\left[(m(\mathbf{r}, \tilde{t}))^2\right] d\mathbf{r} = |C|^2 \leq 1 \tag{5.26}$$

When we had a fairly insensitive medium like film, we would not see the first term since it is proportional to $\bar{k}_{\tilde{t}}$. This represents the shot noise or quantum noise. This has to be large to expose the film, hence the dominant term is the $\bar{k}_{\tilde{t}}^2$ term. If we look at a row of pixels across an image, it appears as in Figure 4.7. This is the image $m(\mathbf{r}_j, \tilde{t})$ riding on a constant level characterized by the Delta function. By construction $0 \leq |C|^2 \leq 1$. This is obvious since intensity cannot go negative. If additive background noise is present, it contributes to both the constant level and the shot noise, but not the image. Remember an observer can only distinguish greater than about 2% ripple in the intensity before all contrast is lost. To compensate for this, the eye has a pupil which opens in dim light and closes when the light is intense (a built in AGC). In addition, the retina has the ability to desensitize with intensity. The difference in intensity between the Sun and the full Moon is 56 dB, yet the eye has almost full contrast at both extremes. Since most photography was taken in bright light (or with flash bulbs) it was always the case that the background was greater than the signal (or was the signal) to expose the film. In fact the correct exposure for a photographic print would be on the steep portion of the gamma curve. Consequently, if the observer were not able to see the shot noise, and since the exposure is proportional to intensity, the natural definition would be

$$\text{Contrast} = \frac{C\bar{k}_{\tilde{i}}}{\bar{k}_{\tilde{i}b} + \bar{k}_{\tilde{i}}} \approx \frac{C\bar{k}_{\tilde{i}}}{\bar{k}_{\tilde{i}b}} \quad : \quad \bar{k}_{\tilde{i}b} >> \bar{k}_{\tilde{i}} \tag{5.27}$$

to account for the average level of the contribution of the image. Along comes new technology which can filter out the DC term. Now the true limit, signal-to-noise ratio, appears, Eqs. (4.44), (4.45). What do we do? We suggest that we partition the signal-to-noise ratio equation into a product of three terms:

$$\text{SNR} = \frac{\eta C_0^2 \bar{P}_s^2}{(\bar{P}_s + \bar{P}_b + \bar{P}_d) 2hfB_e} = C_0 \frac{1}{\left[1 + \left(\frac{\bar{P}_b + \bar{P}_d}{\bar{P}_s}\right)\right]} \left(\frac{\eta C_0 \bar{P}_s \tau_e}{hf}\right) : \quad \tau_e = \frac{1}{2B_e}$$

$$\tag{5.28}$$

The first is the inherent contrast of the scene as defined by C_0. The second term is the loss in contrast (contrast attenuation) as we propagate, and finally the third term is the quantum-limited signal-to-noise ratio that would exist without additive noise, at the source, that is the inherent signal-to-noise ratio. The product of the inherent contrast and the loss in contrast is the apparent contrast. The appearance of inherent contrast in the term for quantum-limited signal-to-noise ratio merely states that not all the signal power contains information. What we have accomplished is to anchor the previous work on imagery and contrast into a rigorous information context. In this context we see that there is no visibility limit to what can be observed, only a loss in signal-to-noise ratio. Thus visibility is merely a consequence of the frailty of the human eye. The trick becomes extracting the image, albeit with a deteriorated signal-to-noise ratio, out of the measured return.

Notice also (Example 4.4) that the signal-to-noise ratio is proportional to the square of C, hence it takes a 6 dB change in the SNR to measure a 3 dB change in inherent contrast. Since a digital conversion of the current gets 6 dB of SNR for every bit, at low levels of contrast we can only measure a 3 dB contrast variation. Consequently, if 2% contrast variation is the visibility limit (−17 dB), it would take 5.66 bits of resolution just to reach zero visibility with zero SNR. To reach sub-visibility of −10 dB (with SNR = 0 dB) would require another 3.33 bits or a total of 9 bits. An ADC with 12–15 effective bits should be able to give a fairly good image (~10 dB SNR) at −10 dB visibility (Figure 5.3). This would be a daylight scene in the MWIR with a 2 μm bandwidth (10 cm aperture and 10 cm resolution at 3000 meters, for both 1 and 60 Hz electrical bandwidth). Rigorously we have

$$10 \log \text{SNR}_b = 10 \log C_0 - 10 \log\left[1 + \left(\frac{\bar{P}_b + \bar{P}_d}{\bar{P}_s}\right)\right] + 10 \log\left(\frac{\eta C_0 \bar{P}_s \tau_e}{hf}\right) \tag{5.29}$$

The theory developed by Duntley assumed we could process the fields directly. However, all optical fields eventually must deal with the quantum effect if they are to be processed for viewing. That is the main difference between the development of contrast in Chapter 4 and past treatments. In addition, we have derived the structure of the image, $m(x,y)$ and its information content, directly from the detection process, and in a manner to

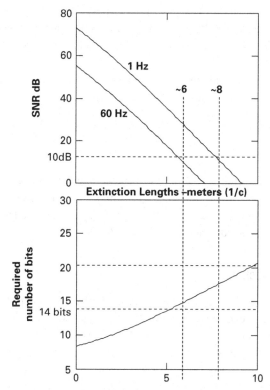

Figure 5.3 Signal-to-noise ratio (SNR) and required number of bits as a function of signal loss extinction length for 1 hertz and 60 hertz frame rates in MWIR.

which signal processing can be applied. Since the range of $m(x,y)$ was $(-1, \infty)$, corresponding to the range presented by Duntley, we chose to average $m^2(x,y)$ over the image to produce C^2 so that the quantum-limited SNR would always show positive. This is a carry-over from communication theory where m is the modulation index. (To reflect the desired interpretation by the eye, $m(x,y)$ should be raised to a level greater than zero.) This is a minor inconvenience since it always appears as $|C|^2$ in the SNR equation, and can be treated as such in the equation for contrast. However, having now made this association we can identify $\bar{P}_b$ with the background radiance. We can also associate $\bar{P}_s$ with both the attenuated signal and the signal-generated clutter.

Notice now that in the absence of background (and detector) noise, the inherent contrast of the object equals C_0 (or $m(x,y)$ if one prefers the signed version over different portions of the image). Also, since this is an integral part of the calculation for SNR, it is more fitting for our purposes to show the loss in contrast

$$\left(\frac{C\bar{P}_s}{(\bar{P}_s + \bar{P}_b + \bar{P}_d)} \right) = \frac{C_0}{1 + \left(\dfrac{\bar{P}_b + \bar{P}_d}{\bar{P}_s} \right)}$$

$$= \frac{\text{Inherent}}{\text{Contrast}} \times \frac{\text{Contrast}}{\text{Attenuation}} = \text{apparent contrast} \qquad (5.30)$$

as diminishing the inherent (quantum-limited) signal-to-noise ratio of the image itself. Other than these points, there are no differences from our treatment and that developed by Duntley. In fact, we benefit greatly from the work done in radiative transport when we consider actual systems.

As a final comment we should point out that for the focal plane array, $h(\mathbf{r})$ includes the effects of the lens and spatial quantization by the array itself.

5.5   Target attenuation

There are two different types of targets that we must consider, passive images and active illumination. We distinguish between the two by assuming that images are formed by passive viewing and reflections created by active interrogation; i.e., radar and lidar systems. Active viewing will be considered in Chapter 7.

Passive imaging from reflectance

When we form a passive image, we must consider several aspects of the problem. First and foremost is obtaining the desired resolution. This sets the pixel field of view (FOV$_{pixel}$), which is either A_d/f_c^2, or λ^2/A_r if we use diffraction-limited optics. If the image is taken during the day, it will be illuminated by the Sun. (An extensive set of curves is presented in Chapter 1.) If the solar radiation is unobscured, and I_{sun} is the spectral irradiance, then the amount of power emitted by the target, with reflection coefficient η_t in the area observed by a pixel at range z, will be $\frac{\eta_t I_{sun} \text{FOV}_{pixel} R^2 \Delta\lambda}{\pi}$ (Eqs. (2.76), (2.80)). Here we have ignored the solar angle, and have assumed a Lambertian surface. Finally at the receiver the optics will subtend a solid angle to the target equal to A_r/R^2. Thus the total received power will be

$$P_s = \frac{\eta_t I_{sun} \text{FOV}_{pixel} \Delta\lambda A_r}{\pi} = \frac{\eta_t I_{sun} A_d \Delta\lambda A_r}{\pi f_c^2}$$

$$= \frac{\eta_t I_{sun} \lambda^2 \Delta\lambda}{\pi} : \quad \text{diffraction-limited} \tag{5.31}$$

In a structured object, each pixel would display a different angle towards the receiver (creating the image $m(r)$), so we have assumed an average cosine of unity. If we are forming an image, any scattered radiation will tend to obscure the image, and hence must be treated as background noise. We distinguish by z the travel in the scattering medium, and R the total range. If the Sun is at an angle φ relative to the surface normal, and if the receiver is at an angle β relative to the surface normal, there will be an additional loss equal to $\cos\varphi \cos\beta$.

Passive imaging by emission

If this was an image taken at night, the power emitted by the target will be determined by its temperature. Depending upon the sensitivity of the sensor, the signal component of this power in the (i, j) pixel will be $\varepsilon(i, j)\Delta P_{BB}(i, j)$ as described in Chapter 4, and the received signal power from that pixel will vary from the average scene temperature by

$$
\varepsilon_{i,j}\Delta P_{BB}(i,j) = \left[\varepsilon_{i,j} \frac{e^{\frac{hc}{kT_{i,j}\lambda}} \frac{(hc)^2}{kT_{i,j}^2}}{\left(e^{\frac{hc}{kT_{i,j}\lambda}}-1\right)^2} \frac{A_r A_d c}{\lambda^6 f_c^2} \right] \Delta\lambda\Delta T
$$

$$
= \left[\varepsilon_{i,j} \frac{e^{\frac{hc}{kT_{i,j}\lambda}} \frac{(hc)^2}{kT_{i,j}^2}}{\left(e^{\frac{hc}{kT_{i,j}\lambda}}-1\right)^2} \frac{c}{\lambda^4} \right] \Delta\lambda\Delta T : \quad \text{diffraction-limited} \qquad (5.32)
$$

The attenuated signal will then be $\varepsilon_{i,j}\Delta P_{BB}(i,j)e^{-cz}$, while the background will attenuate as $\varepsilon_{i,j}P_{BB}(i,j)e^{-kz}$. If the illumination is solar, then both reflected and emitted radiation will emerge from the target. The larger one will dominate, with the scene being reflective during the day and transition to emissive during the evening.

5.6 Additive background noise – reflective background

In the simplest case of additive background noise we have the Sun illuminating the scattering medium Figure 5.2. This could be the sky, haze, fog or a cloud. In the first three cases, the receiver would not be looking directly at the Sun, and in the worst case the radiance from the scattering medium would fill the field of view of the receiver. We would characterize the medium by a diffuse reflection coefficient, ρ_0. The amount of noise entering a single pixel would then equal, P_b, where

$$
P_b = \frac{\rho_0 I_{sun} A_d A_r \Delta\lambda}{\pi f_c^2}
$$

$$
= \frac{\rho_0 I_{sun}\lambda^2 \Delta\lambda}{\pi} : \quad \text{diffraction-limited} \qquad (5.33)
$$

The signal on the other hand will also be illuminated by the Sun, but would experience an attenuation of e^{-cz} while propagating to the receiver. We also allow for the possibility of a finite fog with the receiver located a distance $R - z$ from the boundary. The signal-to-noise ratio becomes

$$
\text{SNR} = \frac{\eta(C_0 P_s e^{-cz})^2 \tau_e}{(P_s e^{-cz} + P_b' + P_d)hf} = \left\{ \frac{C_0}{[1 + e^{cz}(P_b' + P_d)]} \right\} \frac{\eta C_0 P_s \tau_e}{hf} \qquad (5.34)
$$

where we have identified the three terms: inherent contrast, contrast attenuation and inherent quantum-limited signal-to-noise ratio defined at the source. The maximum value for P_b' occurs when the target is totally obscured, for which case Eq. (5.32) applies. If the target is not obscured, $P_b' < P_b$. P_s becomes

$$P_s = \frac{\eta_t I_{sun} A_d A_r \Delta \lambda}{\pi f_c^2}$$

$$= \frac{\eta_t I_{sun} \lambda^2 \Delta \lambda}{\pi} \quad : \quad \text{diffraction-limited} \qquad (5.35)$$

Notice that the signal-to-noise ratio is directly proportional to the integration time τ_e. This implies that for a static image the signal-to-noise ratio can be increased and hence the apparent contrast improved. This frame integration will add to the dynamic range of the digital system. This improvement can continue until inherent limitations to contrast improvement arise, such as calibration. In Duntley's terminology this would be called the apparent signal-to-noise ratio, whereas what we have called the quantum-limited signal-to-noise ratio would be called the inherent signal-to-noise ratio.

5.7   Additive background noise – emissive background

If this is a nighttime scenario, the noise would be the black-body emission from the target, Eq. (4.15) ($2B_0 = c\Delta\lambda/\lambda^2$)

$$P_{BB} = \left(\frac{hc}{e^{\frac{hc}{kT_{i,j}\lambda}} - 1} \right) \frac{A_d A_r c \Delta \lambda}{\lambda^5 f_c^2}$$

$$= \left(\frac{hc}{e^{\frac{hc}{kT_{i,j}\lambda}} - 1} \right) \frac{c\Delta\lambda}{\lambda^3} \quad : \quad \text{diffraction-limited} \qquad (5.36)$$

(We distinguish between the electrical bandwidth $\tau_e = \frac{1}{2B_e}$, which is the frame rate, and the optical bandwidth $2B_0 = c\Delta\lambda/\lambda^2$, which is related to the optical filter line width.) The emissive background noise will only diminish as e^{-kz} assuming the fog is extended. Hence the signal-to-noise ratio will be

$$SNR_{BB} = \varepsilon_{i,j}\, \eta \left(\frac{\left(e^{\frac{hc}{kT_{i,j}\lambda}} \frac{(hc)}{kT_{i,j}^2} \right)^2}{\left(e^{\frac{hc}{kT_{i,j}\lambda}} - 1 \right)^3} \right) \frac{A_r A_d \Delta \lambda \Delta T^2 e^{-2cz+kz} \tau_e}{\lambda^7 f_c^2}$$

$$\varepsilon_{i,j}\, \eta \left(\frac{\left(e^{\frac{hc}{kT_{i,j}\lambda}} \frac{(hc)}{kT_{i,j}^2} \right)^2}{\left(e^{\frac{hc}{kT_{i,j}\lambda}} - 1 \right)^3} \right) \frac{\Delta \lambda \Delta T^2 \tau_e e^{-2cz+kz}}{\lambda^5} \quad : \quad \text{diffraction-limited} \qquad (5.37)$$

(Eqs (4.51) and (4.52)). If this is a day/night transition scenario then we would add the reflective terms with the appropriate input for the Sun.

Example 5.2 What happens to sunlight entering the top of a cloud? If we were under a cloud looking up at a target beneath the cloud, the radiation from the Sun exiting the cloud bottom would be attenuated. In the atmosphere water deposits on aerosols and creates near perfect spheres, characterized by Figure 5.1a. However, because there are so many scatterings occurring in the clouds, the light is totally diffused. In addition, there is a slight amount of absorption because of the finite complex index of water at visible frequencies. The albedo for single scattering, defined as the ratio $\omega = b/c$, has been empirically shown to be in the range of 0.9–0.999 [19, pp. 334–335]. Thus light enters a cloud and reverberates until it either leaks out the bottom, or back out the top. The thicker the cloud, the darker it looks from below (because more is backing up through the top). However, it always looks white looking down at the top. A model for cloud transmission has been estimated as T_c [21], where

$$T_c \approx \frac{1.69 + B(\phi)}{1.42 + \tau_d} \tag{5.38}$$

with $B(\phi)$ a sixth-order polynomial in φ, and the diffusion thickness $\tau_d = \tau(1-<\cos\theta>) \approx \tau(\theta_{rms}^2/2)$, with T_p the physical thickness of the cloud, $\tau = cT_p$ the optical thickness, and θ_{rms} the rms scattering angle of the cloud particulates [20, 22]. (A more detailed discussion of cloud propagation will be given in Chapter 11.)

Looking up at the cloud, the receiver would experience this attenuation if it observed an infinite cloud bottom. What is observed is a Lambertian surface, with a receiver field of view, or

$$P_b = \frac{T_c I_{sun} A_d A_r \Delta\lambda}{\pi f_c^2}$$

$$= \frac{T_c I_{sun} \lambda^2 \Delta\lambda}{\pi} \quad : \quad \text{diffraction-limited} \tag{5.39}$$

The signal on the other hand is illuminated by the sky and cloud bottom (Figure 1.5). The size of the cloud and the Sun angle would determine whether the sky is directly illuminated by the Sun, or is attenuated by T_c.

5.8 Measured data

A few examples of the scenario in Figure 5.3 are shown in Figures 5.4 through 5.12. All the pictures presented in this chapter (including the web site) were provided by David Buck from SPAWAR in San Diego, who has been collecting and processing these data

Figure 5.4 Images of experimental site where image data was taken in fog.

since 2005.[2] The primary site for the collections is shown in Figure 5.4. It is located off the tip of Point Loma, CA, and faces west, from where fog and haze roll in. In this section the processing used is what we refer to as ad hoc processing. Three steps are involved. First is a bias removal to eliminate the constant term. Next is an assessment of the histogram, which for the low-visibility scenes consists of extracting that portion of the histogram lying between the plus and minus three sigma points. Finally the histogram is linearly stretched over the entire dynamic range. In Figure 5.5, taken from point A in Figure 5.4, one can see the Coronado islands off Mexico located south of Point Loma, a distance of 16.7 miles. Although multiple cameras at multiple wavelengths were used, these pictures are from two identical cameras with two different band-pass filters, one covering 450–700 nm and the other 700–1000 nm. Notice that neither target was visible with the naked eye. By removing the DC term from the data and linearly stretching the data from zero to maximum we obtained the result in Figure 5.6. Notice also that while there is less energy radiated from the Sun in the near IR band, that the camera had a greater return with higher contrast. Furthermore there was enough correlation between the two bands to increase the contrast by a factor of two, by subtracting the correlated noise (Figure 5.7). These data are not calibrated since we

---

[2] Private communications.

Clear and Hazy Images of Los Coronados Island

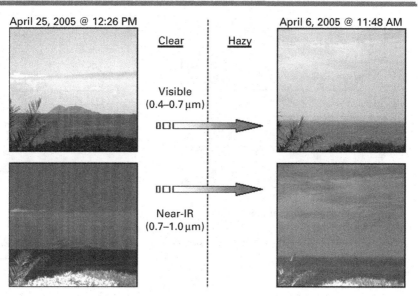

Figure 5.5

Weighted-Difference of a Single Pair of Frames

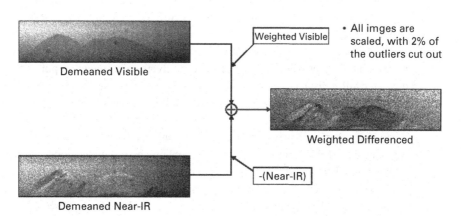

Figure 5.6

don't know the extent of the haze or the contrast at the source. The cameras used had 12 bits of resolution with no AGC. Suffice it to say that of all the bands covered only the LWIR camera was blank, indicating that there was enough water along the path to extinguish the return.

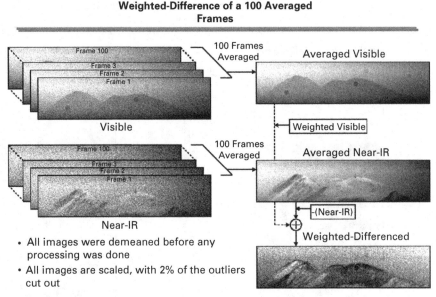

Figure 5.7

A second set of data is shown in Figure 5.8, taken from the same site. The ship is approximately 2 miles off shore. A dynamic version of this was also taken and is presented at the web site www.cambridge.org/karp-stotts as Figure Web-1. Although the pictures were taken in real time, the processing was performed frame by frame.

When the same scenario is applied in Figure 5.3 to measurements in the fog, the site at point B looking down the hill to site A was used. This is required because of the difficulty in both taking and calibrating measurements in the fog. When the fog rolls in, the first few meters are evaporated by the warm earth. Consequently, it was impossible to make measurements over a horizontal path. The receivers were moved to point B and looked down the hill, a distance of over 400 meters. This enabled pictures to be taken through the fog, but calibration was difficult. Using two lights at known locations and measuring the difference in intensity (at 10 μm) was used to estimate c at the LWIR wavelength. Since the uniformity of the fog wasn't known it was only qualitative. In Figure 5.9 quadrants of the same data (LWIR) are shown with different processing. In the upper left quadrant are unprocessed data. In the upper right-hand quadrant commercial processing routines were used, where the data were tiled and histogram equalization was used. This effectively removes the DC and equalizes the quantiles in the histogram. Since the commercial routines did not have a linear stretching routine to apply to the tiles, one was written. In the lower left-hand quadrant an early version of the filter was used and in the lower right-hand quadrant the ALS filter [24] was used, which normalized the histogram and then linearly stretched it over the full dynamic range. While not a linear transformation, it is an affine transformation and preserves most linear properties. All the processing was done frame by frame. A dynamic version of these data can also be found on the web as

Figure 5.8 Long range imaging of a ship in haze.

Figure Web-2. A Similar result was obtained in the MWIR band, Web-3. No return was discernible in the visible, NIR, or SWIR bands (Figure Web-4). A conclusion that can be drawn from these measurements is that for shorter distances up to a few kilometers, the LWIR and MWIR work best in fog because of the longer scattering lengths (smaller values of b). If the paths are too long, even in light haze the bands with low water absorption work best. In the rain there are mixed results because the rain tends to cool the targets and minimize the inherent contrast.

If one looks at the Sun as a source and a cloud as an obstacle, one should be able to look through a thin to light cloud. To test this out direct measurements of the Sun in the visible were made. No success was achieved since it was never certain where the Sun was. Pictures of the Sun were taken with a lot of glow around it (Figure Web-5). Because of the extensive saturation, the exposure was shortened until the limit of the camera was reached. There were still saturated pixels, but a lot of the glow was processed away. More success was obtained at the LWIR. Here a lower solar irradiance and a longer scattering length exist (Figure Web-6). Notice that the disc was successfully imaged,

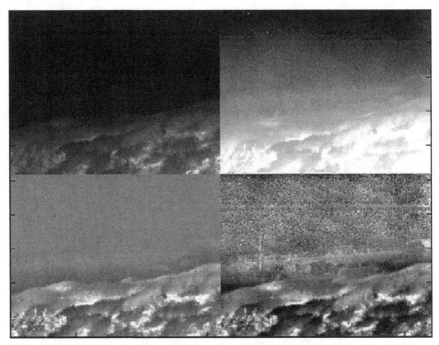

Figure 5.9 Sample LWIR data that is processed differently in each quadrant.

supporting the argument about the unscattered radiation preserving the image. This particular LWIR camera was an uncooled bolometer. Its advantage was a 14-bit dynamic range.

Although there are people working in the camera companies using what is referred to as LAP (local area processing) and homomorphic filtering, what is found in the literature are papers that deal with using these techniques to enhance blurred images, to obtain super resolution [23], or to enhance shadow regions. They are basically using the tiling programs in signal processors, together with company proprietary algorithms. Photographers also use autocontrast and other features of signal processors, but these adjust the mean and balance of the pictures to make them more pleasing to the eye. However, no publications or commercial instruments are on sale to perform sub-visibility imaging. (TruView is a commercial product employing homomorphic filtering whose product assumption does not apply to the scattering channel, hence is not applicable.)

5.9   Other applications

At our website are displayed other interesting applications that were addressed. On the static side measurements have been made off the pier at Ocean Beach in California. It was a simple test to see if the bottom could be observed. It was shallow littoral water but it was possible to image the bottom (Figure Web-7).

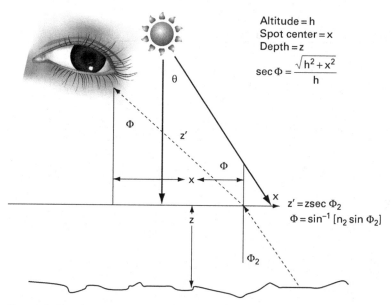

Figure 5.10 Imaging geometry with key parameters defined.

Example 5.3 What can be seen looking down into the water? Consider Figure 5.10, where a view of the ocean bottom is desired. The solar constant at the top of the atmosphere averages about 1366 W/m², which is the integral over the solar spectrum. The spectral irradiance (W/m² · Å), where Å denotes angstroms (0.1 nm), is shown in Figure 1.4 as a function of the zenith angle. In the visible, which goes from 0.45 to 0.65 μm, there are few absorption lines, and the value 0.15 m² · Å is customarily used. The water, however, has different characteristics (Figure 5.11). We will consider Jerlov Type II, which is a common coastal water. When the Sun is the illuminator, it is also the primary source of noise. The signal-to-noise ratio will become

$$\text{SNR} \approx \frac{\left[I_{target} A_{aperture} \right]^2}{I_{noise} A_{aperture} 2hfB} \tag{5.40}$$

This is because the signal will be characterized by the mean of the characteristic to be observed, whereas the noise will only be a variance. Unfortunately,

$$\text{Contrast} \approx \frac{I_{target} A_{aperture}}{I_{noise} A_{aperture}} \tag{5.41}$$

so that it is possible to have a condition where the target (illuminated by the Sun) cannot be seen by the eye, $C < 0.02$, but there is plenty of SNR. By proper signal detection and processing this can be circumvented and detection can be correctly made on the target.

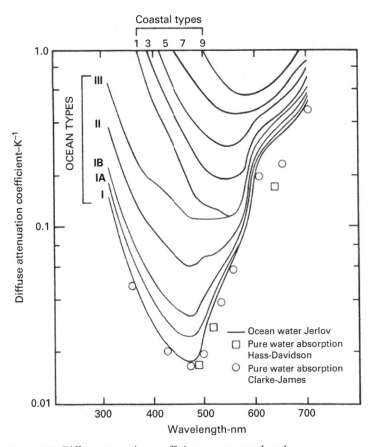

Figure 5.11 Diffuse attenuation coefficient versus wavelength.

The statistics during the daytime will obey the central limit and be Gaussian. We compute the power as follows: if the solar spectral irradiance is I_s, the spectral irradiance on the ocean surface is $I_s \cos \theta$ where θ is the zenith angle the Sun makes. The illumination of the target at a depth z is $I_s \cos \theta\, e^{-kz}$, where k is the diffuse attenuation coefficient and accounts for scattered and absorbed power. However, at $t = 0$, when the target is illuminated, all scattered and absorbed radiation detracts from the image. The unscattered portion of the image is attenuated as e^{-cz} where $c = a + b$, the extinction coefficient, is the sum of the absorption and scattering coefficients with $c \geq k \geq a$. Therefore, the two-way loss of the Sun's energy becomes $I_s \cos \theta\, e^{-kz}\, e^{-cz}$. Unfortunately, the Sun reflects off the ocean surface, so that is where the greatest source of noise is. This diffuse reflection coefficient ρ_0 (Figure 5.12) is approximately 3% at these wavelengths. What the receiver at altitude sees for noise is

$$\rho_0 I_s \cos\theta \, \frac{\cos\Phi}{\pi} \left[A_{receiver} \Big/ (h \sec \Phi)^2 \right] \qquad (5.42)$$

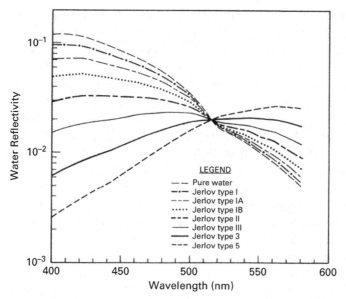

Figure 5.12 Water reflectivity ρ as a function of wavelength for Jerlov water types I, II, III, 3, and 5.

For the signal we assume it will be refracted as $\cos \Phi$. Thus for the received signal we estimate the return per square meter to be

$$\eta I_s \cos \theta e^{-kz} e^{-cz} \frac{\cos \Phi}{\pi} \left[A_{receiver} \Big/ (h \sec \Phi + z \sec \Phi_2)^2 \right] \qquad (5.43)$$

where η is the reflection coefficient of the target, which we take to be 5%. The signal-to-noise ratio then becomes

$$\text{SNR} \approx \frac{[I_{target} A_{aperture}]^2}{I_{noise} A_{aperture}} = \frac{\left\{ \eta I_s \cos \theta e^{-kz} e^{-cz} \dfrac{\cos \Phi}{\pi} \left[A_r \Big/ (h \sec \Phi + z \sec \Phi_2)^2 \right] \right\}^2 N \tau_e}{\rho_0 \eta I_s \cos \theta \dfrac{\cos \Phi}{\pi} \left[A_r \Big/ (h \sec \Phi)_2 \right] hf}$$

$$= \frac{\dfrac{1}{\rho_0 \pi} \cos \theta e^{-kz} e^{-cz} \cos \Phi \left[A_r (h \sec \Phi)^2 \Big/ (h \sec \Phi + z \sec \Phi_2)^4 \right] \eta I_s N \tau_e}{hf} \quad : N \tau_e = 1/2 B_e$$

$$(5.44)$$

and the contrast at the water surface becomes

$$\text{Contrast} = \left\{ \frac{\eta \cos \theta}{\rho_0 \pi} e^{-kz} e^{-cz} \right\} \qquad (5.45)$$

Assuming that an SNR of 10 dB is sufficient for recognition, 12–14 effective bits are needed.

The ALS filter has also been used to enhance the soft tissue images in X-rays (Figure Web-8). Another interesting static image was that of a fire in Rancho Bernardo (Figure Web-9). Because of the smoke no visible images were possible. However, both MWIR and LWIR could penetrate it. Notice that the image was greatly enhanced by the ALS filter; more so for the MWIR because of the heat of the fire. This might have particular application in mines when there are accidents and the visibility is practically zero because of the coal dust. If the temperature is too uniform for thermal imaging, low-power LWIR illumination could be used to create the reflection. Similar results were obtained in measurement of brownout (Figure Web-10). This is the term assigned to the dust that is created by helicopters in sandy areas. Sand has a near-zero lost tangent at the LWIR, which makes it ideally suited for imaging. The last application to be discussed applies to aircraft landing under Instrument meteorological conditions. This is where the runway cannot be seen by the aircraft until it is close to landing. No flights were made under those conditions but one can see the runway in a flight that went through a fog bank (Figure Web-11). The implication is that camera on the aircraft may allow the runway environment can be seen before the approach minimum (ceiling).

5.10   Spatial filtering

In all the images shown in section 5.7, two ad hoc rules have been used, based upon earlier work [24]. The first was to remove the DC bias in the image. The second was to develop metrics to establish the image in the histogram, eliminate the outliers and linearly stretch the low-contrast portion of the histogram over the entire dynamic range. At this point no attempt has been made to develop spectral filters for the image to increase the signal-to-noise ratio. However, there is detailed knowledge of the image. For example $H(f_x, f_y)$ in Eq. (5.24) is the deterministic optical transfer function of the entire system. Since the electron count in each pixel will be high, $H(f_x, f_y)$ can be pre-measured and used for pre-whitening in a variety of ways. Also the techniques of compressive sampling can be used to enhance the signal-to-noise ratio of the image. Since both the signal and noise approach Gaussian statistics, SNR estimators can be constructed to sample the spatial image and aid in all aspects of processing. These techniques are illustrated in Figure 5.13.

Example 5.4 What is the reconstruction of the image? From Eqs. (2.52), (2.54), we see that the mutual coherence function is convolved with the transfer function of the channel, which is a multiplication in the Fourier transform domain. Similarly we see that there is a second convolution of this result with the optical transfer function (OTF) of the optics, which again multiplies in the Fourier transform domain. Thus we see that the mutual coherence function of the signal is filtered first by the channel and next by the receiving optics. Finally, the image is sampled by the two-dimensional focal plane array [25, p. 40].

$$\iint I_a(\xi, \eta)\delta(x - \xi)\delta(y - \eta)d\xi d\eta = I_a(x, y) \tag{5.46}$$

Energy Spectrum of Image

$$\bar{k}_s^2 S_{s,\tilde{i}}(k_x, k_y)$$

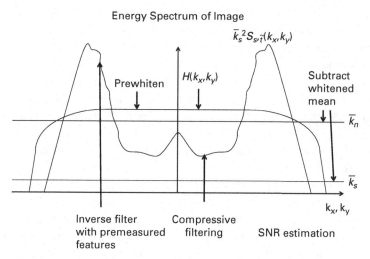

Prewhiten

$H(k_x, k_y)$

Subtract
whitened
mean

$\bar{k}_n$

$\bar{k}_s$

k_x, k_y

Inverse filter
with premeasured
features

Compressive
filtering

SNR estimation

Figure 5.13

If we assume that the focal plane array has square pixels $2a$ by $2a$ on $2a$ centers, then the reconstruction of the image takes the form [25, p. 128],

$$I_d'(x,y) = \sum_{n=-\infty}^{\infty} \sum_{m=-\infty}^{\infty} \left(\frac{\pi}{a}\right)^2 I_d\left(\frac{n\pi}{a}, \frac{m\pi}{a}\right) \left[\frac{\sin a(x-n\pi/a)}{a(x-n\pi/a)}\right] \left[\frac{\sin a(x-m\pi/a)}{a(x-m\pi/a)}\right]$$

(5.47)

As long as the spatial sampling rate $1/a$ is greater than the highest spatial frequency in $I_d(x,y)$, the image will be reproduced faithfully. The most important element in filtering the two-dimensional signal, is being able to identify the spatial frequencies that have sufficiently high signal-to-noise ratios. Consequently, one needs a good signal-to-noise ratio estimator.

The photo-detector (pixel) will either be a form of a solid-state device that has "wells" which collect electrons, or some form of a bolometer that responds to heat. In either case there will be hundreds of thousands, even millions of electrons collected in a time-integrated sample in each pixel. The current in a pixel will therefore be Gaussian with less than 0.1% deviation [26, p. 116]. The Fourier transform can be written as the sum of in-phase and quadrature-phase components each modulated by a Gaussian random variable, and each with the same mean and variance. We thus establish that the distribution of the spatial spectrum will also be Gaussian. We normally would be measuring the envelope of the spectrum. However, the envelope distribution would be Rayleigh distributed for the noise, Ricean distributed for signal plus noise and we don't have an estimator for the signal-to-noise ratio. As the real and quadrature components are both Gaussian, Gagliardi [27] shows that we need address only the real part of the spectrum with a similar estimate for the signal-to-noise ratio of a single frequency element in a local region in the f_x, f_y plane.

From Eq. (5.24) it can be seen that the system function $H(f_x, f_y)$, which multiplies the entire two-dimensional spectrum, includes the channel, the optics, the photo-detector response, the calibration and the two-dimensional quantization by the focal plane itself. The first two are embedded in the intensity incident on the focal plane, while the latter three are contributed by the focal plane. However, while there will be some degree of aging, this is not a random function and it can be premeasured and stored. It is necessary to measure extremely low values of contrast, and consequently it is important to calibrate and linearize the detector response over the entire dynamic range first, so that response linearity is not intensity-dependent. At least a 16-point calibration from a uniform source is recommended. Then we should compute the spectrum, remove the DC term, and preferably pre-whiten it with the inverse of $H(f_x, f_y)$. In lieu of that one could eliminate the very low-frequency content.

The spectrum should be sampled over a homogeneous region to ensure that the signal is also homogeneous. This is true if the size of each sample is small, so the spectral content doesn't vary. As the size of the sample is increased, the estimate first stabilizes while the size of the local area increases, until it diverges because the spectrum changes. A happy medium should be found. Sliding a three-by-three to five-by-five averager over the spectrum should suffice. This would allow us to develop an SNR map of the spectrum. Having this we can eliminate those portions with low SNR by setting a threshold. Finally the inverse transform of the spectrum should be taken, tiled with the outliers removed, and then one should linearly stretch the histogram over the full dynamic range, merge the tiles and reconstruct the image.

Unfortunately, at the time of this printing a signal-to-noise ratio estimator has not been established, although several options appear available [28–32]. Figures Web-12–Web-15 on the web page are reserved for when these results are available.

5.11   Summary

In this chapter we have brought together all the relevant work that we feel contributes to sub-visibility imaging. Although we have been directly addressing this area for over 5 years, we have not been able to find other published work in the field. There appear to be several niche areas where these results can be usefully applied. To that end we hope this chapter will be helpful.

References

1. World Meteorological Organization. *Guide to Meteorological Instruments and Methods of Observation*, Part I, Ch. 9, preliminary 7th edn, WMO-No.8 (2006).
2. N. Pogson. Magnitudes of thirty-six of the minor planets for the first day of each month of the year 1857. *Monthly Notices of the Royal Astronomical Society*, **17** (1856), pp. 12–15.
3. H. R. Blackwell. Contrast thresholds of the human eye. *Journal of the Optical Society of America*, **36**(11) (1946), pp. 624–643.
4. J. Thaung, C. Beckman, M. Abrahamsson and J. Sjostrand. The light scattering factor. *Investigative Ophthalmology & Visual Science*, **36**(11) (1995), pp. 2313–2317.

5. G. Mie. Beiträge zur Optik trüber Medien, speziell kolloidaler Metallösungen. *Leipzig Annalen der Physik*, **330** (1908), pp. 377–445.

6. H. C. van de Hulst. *Light Scattering by Small Particles*. New York, John Wiley and Sons (1957) (Dover, New York (1981), p. 176).

7. H. C. van de Hulst. *Multiple Light Scattering*, Vol 1. Academic Press, New York (1980).

8. M. Born and E. Wolf. *Principles of Optics: Electromagnetic Theory of Propagation, Interference and Diffraction of Light*, 7th edn. Cambridge University Press, Cambridge (1999).

9. D. Diermendjian. *Electromagnetic Scattering on Spherical Polydisperion*. Elsevier, New York (1969).

10. M. Kerker. *The Scattering of Light and other Electromagnetic Radiation*. Academic Press, New York (1969).

11. J. D. Jackson. *Classical Electrodynamics*, Ch. 6. John Wiley and Sons, New York (1975), pp. 209–254.

12. E. J. McCartney. *Optics of the Atmosphere*, Ch. 4. John Wiley and Sons, New York (1976), pp. 176–215.

13. N. Hojerslev. A spectral light absorption meter for measurements in the sea. *Limnology and Oceanography*, **20**(6) (1975), pp. 1024–1034.

14. E. A. Bucher. Computer simulation of light pulse propagation for communication through thick clouds. *Applied Optics*, **12** (1973), 2391–2400.

15. S. Q. Duntley. The reduction of apparent contrast by the atmosphere. *Journal of the Optical Society of America*, **38** (1948), pp. 179–187.

16. S. Q. Duntley, A. R. Boileau and R. W. Preisendorfer. Image transmission by the troposphere I. *Journal of the Optical Society of America*, **47**(6) (1957), pp. 499–506.

17. S. Q. Duntley. Underwater lighting by submerged lasers. UCSD, SIO ref. 71–1, June 1971.

18. N. G. Jerlov. *Marine Optics*, 2nd edn. Elsevier, New York (1976), p. 157.

19. S. Karp, R. M. Gagliardi, S. E. Moran and L. B. Stotts. *Optical Channels*. Plenum Press, New York (1988).

20. R. F. Lutomirski and H. T. Yura. Propagation of an optical beam in an inhomogeneous medium. *Applied Optics*, **10** (1971), p. 1954.

21. R. E. Danielson, D. R. Moore and H. C. Van de Hultz. The transfer of visible radiation through clouds. *Journal of the Atmospheric Sciences*, **26** (1969), 1078–1087.

22. H. M. Heggestad. Optical communications through multiple scattering media. MIT, RLE Technical Report 472, November 1968.

23. G. C. Holst, E. Cloud, H. Lee, T. Pace, D. Manville and J. Puritz. Super-resolution reconstruction and local area processing (Proceedings Paper). *Proceedings of the SPIE*, **6543** (2007).

24. D. L. Buck *et al.* Patent No. US 8,023,760 B1, date of patent: September 20, 2011.

25. A. Papoulis. *Systems and Transforms with Applications to Optics*. McGraw-Hill, New York (1968).

26. R. M. Gagliardi and S. Karp. *Optical Communications*. Wiley Interscience, New York (1976).

27. R. M. Gagliardi. *Introduction to Communications Engineering*. Wiley Interscience, New York (1978), pp. 495–490.

28. W. Sealy Gosset (Student). The probable error of a mean. *Biometrika*, **6**(1) (1908), pp. 1–25. Probable error of a correlation coefficient. *Biometrika*, **6**(2/3) (1908), pp. 302–310.

29. A. M. Mood and F. A. Graybill. *Introduction to the Theory of Statistics*, 2nd edn. McGraw-Hill, New York (1963), pp. 233, 242.

30. J. S. Bendat and A. G. Piersol. *Random Data, Analysis and Measurement Procedures*, 2nd edn. Wiley Interscience, New York (1986), pp. 80, 525.

31. W. G. Bulgren. On representations of the doubly non-central F distribution. *Journal of the American Statistical Association*, **66** (1971), p. 184.

32. H. Scheffé. *The Analysis of Variance*. Wiley, New York (1959) pp. 135, 415.

33. S. Karp. Statistical properties of ensembles of classical wave packets. *Journal of the Optical Society of America*, **65**(4) (1975), p. 421.

34. R. J. Morgan, R. Hutchin and S. Karp. Direct observation of excess radiation shot noise in multimode chaotic light. DARPA Contract Nos. N00014-83-C-0507 and N66867-85-C-1038 (also PACS No. 42.50+q, 32.80.Fb, 03.65.Sq).

6 Signal modulation schemes in optical communications

Communications systems are designed to send information from one point to another in the face of corrupting noise, signal loss and other degrading effects. Because these effects are statistical in nature, the field of signal detection and estimation was created to provide an analytical means to quantify link performance, establishing the quality of the information transfer. Although communications moved from point-to-point data links to networking among many users in the last decade, link analysis is still important to setting link performance even though networking can help overcome link performance limitations.

The most commonly used parameter for link analysis is the signal-to-noise ratio (SNR), which is normally defined as the ratio of the average signal power to the average noise power at some point in the receiver chain. Although to first order, optical and radio frequency (RF) communications systems operate essentially in the same way, their detection processes are different [1–3].

In an RF receiver, the first detector senses the signal and noise field strengths. The noise input to this detector is usually caused by thermal noise from the antenna and the associated field preamplifier.

On the other hand, optical receivers create electrical current proportional to the received optical power from any external source, e.g., optical source and sunlight, as well as being influenced from the statistical noise generated within the detection process. Optical receivers also do not detect thermal noise sources from the channel as an RF receiver does, but rather derive thermal noise from the electronics of the receiver system. The result is that the electrical SNR for an optical receiver is typically proportional to the square of the incoming optical signal power, and inversely proportional to the sum of the external source and thermal noise powers [1–3]. But this is not all. In the absence of any external source and thermal noise, the received optical signal will not be detected perfectly, but will be subject to a signal variance proportional to the receiver optical signal power. In other words, the detection of an optical signal creates its own noise within the detector. In this case, the electrical SNR is proportional to the received optical power only. This situation is unique to optical systems and has its origins in the quantum nature of light detection. In this case, a high SNR may not yield a low probability of bit error [2, p. 222; 3, Ch. 11].

In this chapter, we will discuss conventional and advanced communications receivers. This will include a discussion of spectral efficiency and the energy per bit per unilateral noise density. These latter aspects help us determine the optimal means for optical communications, especially when erbium-doped fiber amplifiers are involved [4, 5]. One note before we proceed.

In Chapters 3 and 4 where we outlined the fundamental derivations, we tried to keep the notation as simple as possible to avoid distractions. However, when building a system to commercial specifications, it is important to use the language of the modern system engineer. Thus the system engineer has to know what a −55 dBmW receiver sensitivity means when one reads a hardware brochure. Therefore in this and Chapter 8, we have translated the earlier language into a more practical language one uses when building a system. We had considered using a single notation throughout the book, but instead used the philosophy "when in Rome, do as the Romans do". We have included a page of notations to aid the reader in deciphering the text. This also is necessary since letters like 'c' have been extensively used as the speed of light, the volume extinction coefficient, contrast, or an arbitrary constant.

6.1   Spectral efficiency

As noted in Chapter 1, Eq. (1.26) was developed by Claude Shannon to describe the fundamental capacity of any communications system in the presence of noise [6,7]. Today, two key parameters come out of this equation that better quantify the performance of a system; they are the spectral efficiency and energy per bit per unilateral noise density. Using Eq. (1.26), we have

$$C = B \log_2(1 + SNR) = B \log_2\left(1 + \frac{P}{NB}\right) = B \log_2\left(1 + \frac{rE_b}{N_0}\right) \tag{6.1}$$

where

$$r = r(R_b) \equiv \text{spectral efficiency (bps/Hz)}$$

$$= \left(\frac{R_b}{B}\right) = \log_2(M) \tag{6.2}$$

and

$$\frac{E_b}{N_0} = \left(\frac{P}{N R_b}\right) = \log_2(M) \tag{6.3}$$

Figure 6.1 redraws Figure 1.12 in Chapter 1 for $R_b = C$ to provide the spectral efficiency as a function of the energy per bit per unilateral noise density. By using the specific signal-to-noise ratio for various digital modulation schemes, one can quantify the spectral efficiency and E_b/N_0 to determine the optimum means for communicating information. For example, the channel capacity for quantum noise-limited optical communications system can be rewritten as

$$\frac{R_b}{B} = \log_2(1 + SNR) = \log_2\left(1 + \frac{\eta P_s}{2hfB_0}\right) = \log_2\left(1 + \frac{rE_b}{N_0}\right) \tag{6.4}$$

This expression implies E_b/N_0 can be replaced by $n_s/R_b = \left(\frac{\eta P_s}{2hf\,R_b}\right)$, the average number of photo-counts per bit (ppb) after detection.[1] It can be shown that the minimum for most optical

---

[1] In many papers and books, their authors cite units for n_s/R_b as photons per bit. This is not accurate since the photo-detector must convert the incoming light to photo-electrons or photo-counts with conversion efficiency

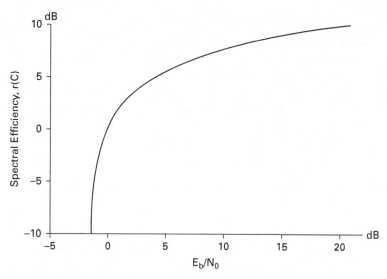

Figure 6.1 Spectral efficiency as a Function of the Energy per Bit per Unilateral Spectral Noise Density under the Shannon Capacity Limit

communications systems where $\eta = 1$ is 1 ppb, with the one exception being homodyne receivers that have a minimum E_b/N_0 of 0.5 ppb [1, Ch. 8].

6.2   Conventional communications receivers

In section 4.6, we introduced the signal-to-noise ratio for digital PPM communications systems. In this section, we broaden that discussion to a more general characterization of optical communications systems. In particular, we will cover advances in these techniques to provide high-bandwidth free-space laser communications similar to what is obtained in fiber optic systems.

In most practical situations historically, the optical SNR is given by the relation[2]

$$\text{SNR} \approx \frac{\left(\dfrac{Gq\eta}{hf}\right)^2 R_L P_s^2}{qB_e G^2\left[\left(\dfrac{q\eta}{hf}\right)(P_s + P_b) + I_D\right]R_L + 4kTB_e} \tag{6.5}$$

where

$G \equiv$ amplifier gain

$q \equiv$ electron charge

η to yield signal and background shot noise. The result is that E_b/N_0 is proportional to the photo-electrons/counts per bit, not photons per bit, even for $\eta = 1$.

[2] Eq. (6.5) focuses on the key noise sources normally associated with conventional systems. Later we will develop the electrical signal-to-noise ratio when erbium-doped fiber amplifiers are used [4], a significant trend in recent optical communications system designs [5–8].

$h \equiv$ Planck's constant $= 6.626 \times 10^{-34}$ joule-seconds
$f \equiv$ frequency of light (Hz)
$\eta \equiv$ detector quantum efficiency (amps/watts)
$R_L \equiv$ resistance in detector load resistor (ohms)
$P_s \equiv$ received optical signal power before the detector (watts)
$P_b \equiv$ received optical background noise power before the detector (watts)
$k \equiv$ Boltzmann's constant $= 1.380\ 6504(24) \times 10^{-23}$ J/K
$T \equiv$ receiver temperature in degrees Kelvin (K)
$B_e \equiv$ detector bandwidth in hertz (Hz) or inverse seconds.

When dark current is negligible, then Eq. (6.5) can be written as

$$\text{SNR} \approx \frac{\left(\frac{\eta}{hf}\right) P_s^2}{(P_s + P_b) B_e} = \frac{\eta}{hf\ B_e} \left(\frac{P_s}{1 + 1/CNR}\right) \tag{6.6}$$

where

$$\text{CNR} \equiv \text{Carrier-to-Noise Ratio} = \frac{P_s}{P_b} \tag{6.7}$$

CNR is the optical signal-to-noise ratio at the input of the first detector of the optical receiver.

When CNR is very large, Eq. (6.6)) reduces to

$$SNR = \frac{\eta P_s}{hf B_e}, \tag{6.8}$$

which represents signal shot-noise-limited, or quantum-limited, communications. This is the limiting case for an optical system when all external and internal noise sources are eliminated. It represents the uncertainty in the number of photo-electrons emitted by the detector [3, p. 178].

On the other hand, when the background noise source is strong, e.g., Sun, sky, Eq. (6.5)) reduces to

$$SNR \approx \frac{\left(\frac{\eta}{hf}\right)^2 P_s^2}{\left(\frac{\eta}{hf}\right) P_b B_e} \tag{6.9a}$$

$$\approx \left(\frac{\eta}{hf\ B_e}\right) \frac{P_s^2}{P_b} \tag{6.9b}$$

Eq. (6.9b) is known as the background-limited signal-to-noise ratio expression and represents the other extreme of optical communications where the background noise source dictates the performance of the receiver. In this situation, narrowing the optical band-pass and the receiver electrical bandwidth of the receiver to the laser signal characteristics will enhance the probability the link will close by moving to a peak power detection scenario. Let us now specify the SNR and probability of bit error for

Table 6.1 Signal-to-noise ratios for baseband and subcarrier direct detection receivers

Baseband direct detection receiver		
	Complete solution for electrical signal-to-noise ratio (SNR) Signal-shot-noise-limited SNR	$\mathrm{SNR} = \dfrac{\left(\frac{Gq\eta}{hf}\right)^2 R_L P_s^2}{qB_e G^2\left[\left(\frac{q\eta}{hf_s}\right)(P_s+P_B)+I_D\right]R_L+4kTB_e}$
		$\mathrm{SNR} \approx \dfrac{\eta P_s}{hf_s\,B_e}$
Subcarrier direct detection receiver		
	Subcarrier SNR	$\mathrm{SNR_{sc}} = \dfrac{\left(\frac{Gq\eta}{hf_s}\right)^2 R_L P_s{}^2}{16qB_{sc}\,G^2\left[\left(\frac{q\eta}{hf_s}\right)\left(\frac{P_s}{2}+P_B\right)+I_D\right]R_L+32kTB_{sc}}$
	Subcarrier signal-shot-noise-limited SNR	$\mathrm{SNR_{sc}} \approx \dfrac{\eta P_s}{8hf_s\,B_{sc}}$
	Signal-shot-noise-limited electrical SNR for amplitude modulation	$\mathrm{SNR} \approx \dfrac{\eta P_s}{4hf_s\,B_e}$
	Electrical SNR for frequency modulation ($\mathrm{SNR_{sc}} \gg 1$)	$\mathrm{SNR} = 3\,(M_{FM})^2\,\dfrac{B_{sc}}{B_e}\,\mathrm{SNR_{SC}}$
		$\approx \tfrac{3}{4}\,(M_{FM})^2\,\dfrac{\eta P_s}{4hf_s\,B_e}$

the various analog and digital communications system designs found in Pratt [3] and Gagliardi and Karp [2]. For consistency, we have made all the equations in the tables in the Pratt notation. Differences with those equations and figures in Chapter 4 will be noted.

Table 6.1 depicts the signal-to-noise ratios that characterize baseband and subcarrier direct detection receiver systems. In this table, f_s is the signal carrier frequency, and B_{sc} is the subcarrier bandwidth given by

$$B_{sc} = 2f_d\left[1 + \frac{1}{M_{FM}}\right] \tag{6.10}$$

where f_d is the frequency deviation of the FM modulator and M_{FM} is the FM modulation index. Comparing the subcarrier direct detection SNR to that of the baseband direction detection system under signal shot-noise operation, the subcarrier approach has a lower SNR by a factor of 4, or 6 dB, than the baseband case.

It is clear the best operation of the above types of communications systems occurs under signal-shot-noise-limited conditions. This means the background and thermal noise sources, as well as any other external or internal noise source, are negligible. Coherent techniques are used to eliminate the influence of these noise sources because the local oscillator at the detection stage overwhelms all non-coherent input signals. There are two varieties of coherent systems; one is called heterodyne and the other homodyne.

Table 6.2 depicts the signal-to-noise ratios that characterize heterodyne detection receiver systems. In this table, f_c is the heterodyne carrier frequency, and B_{IF} is the

Table 6.2 Signal-to-noise ratios for heterodyne detection receivers

Heterodyne detection receiver		
Electrical SNR at the intermediate frequency	$\left[\dfrac{S}{N}\right]_{IF} = \dfrac{G^2\left(\frac{\eta q}{hf_c}\right)^2 P_c P_o R_L}{G^2 q\left[\frac{\eta q}{hf_c}(P_c + P_o + P_B) + I_D\right]B_{IF}R_L + 2kTB_{IF}}$	
Electrical SNR at the intermediate frequency for the local oscillator power much larger than any of the noise sources	$\left[\dfrac{S}{N}\right]_{IF} = \dfrac{\eta P_c}{hf_c B_{IF}}$	
Electrical SNR of a heterodyne receiver for linear envelope detector and a high $[S/N]_{IF}$, or a synchronous detector	$SNR = 2\left[\dfrac{S}{N}\right]_{IF}$	
Electrical SNR of a heterodyne receiver for a strong local oscillator and if the IF filter bandwidth is set to the information bandwidth, B_0	$SNR = \dfrac{2\eta P_c}{hf_c B_0}$	
Electrical SNR for a power detector	$SNR = \dfrac{\eta P_c}{2hf_c B_0}$	

intermediate frequency (IF) bandwidth, B_0 is the information bandwidth, P_c is the received optical power of the carrier signal, P_o is the optical power of the local oscillator and P_B is the received optical power of the background light. The SNR for a heterodyne detection receiver using intensity modulation is one half the signal-to-noise ratio of a signal-shot-noise-limited direct detection receiver.

Example 6.1 For a power detector of dimension d and the laser carrier misaligned by the angle ψ, Pratt showed that the instantaneous IF signal voltage after optical mixing is given by

$$v_{IF} = G\,\frac{\eta q}{hf_c}\,A_c A_o R_L \cos\left[(\omega_0 - \omega_c)t + (\Phi_o - \Phi_c)\right]\frac{\sin(\omega_c d/2v_x)}{(\omega_c d/2v_x)} \qquad (6.11)$$

where

$f_c = {}^{\omega_c}\!/_{2\pi} \equiv$ optical frequency of the laser carrier

$f_0 = {}^{\omega_0}\!/_{2\pi} \equiv$ optical frequency of the laser oscillator
$A_c \equiv$ field amplitude of the laser carrier
$A_o \equiv$ field amplitude of the laser oscillator
$\Phi_c \equiv$ laser carrier phase angle
$\Phi_o \equiv$ laser oscillator phase angle

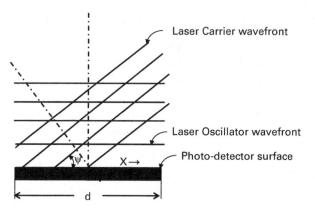

Laser Carrier wavefront

Laser Oscillator wavefront

Photo-detector surface

Figure 6.2 Spatial orientation of carrier and oscillator signal for a collimated optical receiver

$$v_x = c/\sin\psi \tag{6.12}$$

$c \equiv$ speed of light

$d \equiv$ linear dimension of the photo-detector

[3, p. 186]. Figure 6.2 depicts the receiver misalignment situation graphically. Eq. (6.11) implies that the amplitude of the IF voltage is dependent on the misalignment angle ψ.

To keep the signal phase cancellation due to misalignment to 10% or less, we require

$$\frac{\sin\left(\omega_c d / 2 v_x\right)}{\left(\omega_c d / 2 v_x\right)} \leq 0.8 \, \text{radians} \tag{6.13}$$

or

$$\psi \leq \frac{\lambda}{4 d_d} \tag{6.14}$$

Using Eq. (6.14), we find for a 1 μm laser communications systems with a detector of dimension of 1 cm, the angular misalignment must be less than 25 microradians [3, p. 186].

Example 6.2 If the laser carrier beam goes through a converging lens telescope, then the incoming beam will create a focused spot on the photo-detector surface instead of a collimated beam [3, pp. 186–187]. The result is that the misalignment angle is determined by the receiver field of view. The receiver field of view in this case is given by

$$\theta_R \approx 2.44 \, \frac{\lambda \, d_p}{d_D \, d_R} \tag{6.15}$$

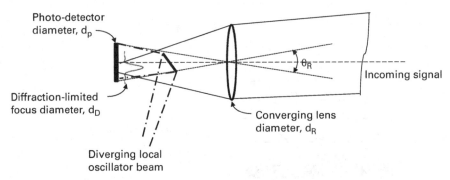

Figure 6.3 Spatial orientation of carrier and oscillator signal for a focusing optical receiver.

where

$d_a \equiv$ photo-detector diameter
$d_R \equiv$ converging lens diameter
$d_D \equiv$ diffraction-limited focus diameter.

Figure 6.3 illustrates this situation, where the laser oscillator beam floods the photo-detector surface uniformly and the incoming beam is focused to a diffraction-limited spot somewhere on the photo-detector surface.

Repeating the above example for a focusing optical system of diameter 10 cm, we find that the diffraction-limited spot has a diameter of 0.01 cm for a 1 μm incoming laser beam, which gives a receiver field of view of 2.44 milliradians. Comparing the two approaches, the latter provides a relaxation in the misalignment requirement by a factor of 100 [3, p. 187]. However, there is one drawback with the converging lens approach. Namely, the photo-mixing of the incoming beam and local oscillator beam only happens within the diffraction-limited spot of the incoming beam. The impact is that the energy of the local oscillator beam outside this spot is absorbed by the photo-detection surface and contributes additional shot noise to the detection process. In other words, the SNR decreases. One can overcome this reduction by employing a focal plane array or image dissection phototube, and only using the photo-current from the areas the incoming beam is focused on [3, p. 187].

Example 6.3 For a circular aperture and detector, the misalignment suppression loss is given by

$$\text{Loss} = \left[2.44 \, \frac{2}{r/R} \int_0^{r/R} J_1(\rho) \, J_0 \left(\frac{4 f_s \, \rho \sin\psi}{d} \right) d\rho \right]^2 \tag{6.16}$$

where

$$R = \frac{2\lambda f_s}{\pi d}$$

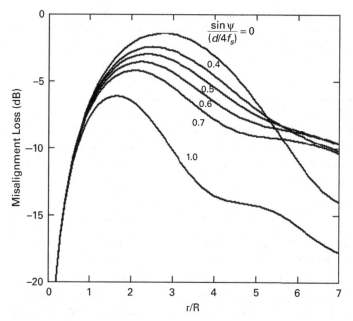

Figure 6.4 Misalignment Loss for a circular aperture as a function of r/R for various values of $\sin \psi / (d/4f_s)$.

Here, r is the detector radius, d is the receiver aperture diameter and ψ is the local oscillator misalignment angle [2, p. 189]. Figure 6.4 shows the misalignment loss as a function of (r/R) for various values of $\sin \psi / (d/4f_s)$.

It is clear from this graph that alignment losses can be reduced by the proper selection of detector size, the latter dependent on the degree of misalignment. In particular, the losses for angles of misalignment less than $\sin^{-1}(d/4f_s)$ are only a few decibels greater than that for a perfectly aligned planed waver ($\psi = 0$), with detectors having $r \approx 3R$. This means that except for the loss factor of using a mismatched plane wave, Eq. (6.15) shows insensitivity to alignment errors for this range of alignment angles. This implies the feasibility of using a single focal plane for heterodyning over an entire array.

The above characterizes an idealized heterodyning using a single spatial mode. When more than one mode is involved, or when conflicting modes between the received and local oscillator fields occur, the above results can be modified [2, pp. 189–194]. Table 6.3 depicts the signal-to-noise ratios that characterize a multi-mode heterodyne detection receiver. In this table, M_R is the number of received modes, M_L is the number of local field modes, A_r is the area of the receiver, a_L is the local field strength and $P_L = |a_L|^2 A_r$.

Comparing with the single-mode heterodyning case, the above SNR is M_R times larger than the single-mode SNR. The system is therefore benefiting from the coherent combining of the signal modes when the frequency and time delay dispersion is negligible. Gagliardi and Karp emphasized that the delay dispersion must be small compared to the heterodyne beat frequency $\omega_{12}/2\pi$. They suggested that equalization techniques could be used to help this problem.

Table 6.3 Signal-to-noise ratios for a multi-mode heterodyne detection receiver

Electrical SNR for $M_R = M_L$,

$$SNR = \frac{\left(\frac{q\eta}{hf_c}\right)^2 A_r P_L / 2 B_0}{1 + 2\left(\frac{\eta}{hf_c}\right)^2 N_{0b} + \left(N_{0c} / \left[M_L \left(\eta / h f_c\right) P_L q^2\right]\right)}$$

Example 6.4 Let us look at what happens when the time dispersion is not negligible. Gagliardi and Karp assumed that they had a set of independent time delays $\{\tau_i\}$, each of which produces a uniform distributed phase over a period, $\omega_{12}/2\pi$ [2, p. 193]. In this situation, the signal power is written as

$$P_s = 2\left[\left(\frac{q\eta}{hf_c}\right) a_L A_r\right]^2 M_L \left(\frac{I_s}{M_R}\right) \tag{6.17}$$

and the resulting SNR is reduced by a factor of $1/M_R$. In other words, there is no longer coherent combining of the heterodyned modes and the multi-mode advantage is lost. Thus, the advantage of heterodyning with many field modes only occurs when the mode delay dispersion is small, or can be reduced. Gagliardi and Karp suggested the delay equalization of the heterodyned modes could be used to trade off field power and bandwidth in a multi-mode detection scheme. Specifically, they proposed that multi-mode direct detection can be achieved over a wide FOV by spatially separating the heterodyned modes and beating against a common local oscillator, i.e., heterodyne and detect each mode separately in a detector array illuminated by one local oscillator beam [2, p 193]. Gagliardi and Karp discussed various techniques for phase probing or other methods for estimating each mode phase, in Chapter 9 of reference [2].

A final note is that non-stationarity of the space channel may seriously hinder the practical implementation of the above types of equalization. The local field still has the requirement for matching and aligning to each field mode, but in addition, now must have enough strength in each mode to overcome the circuit noise of each individual detector. If the field is not properly matched, a power suppression factor must be included for each detector, which may again negate the coherent combing advantage if too severe [3, p. 193].

Finally, the authors noted that delay equalization may have its most practical application in the fiber optic channel, where the fibers tend to produce many signal modes because of real world effects under statistically stationary conditions [3, p. 193]. Heterodyning over an assumed single spatial mode produces power loss because of the unusable excess modes. There one must improve the fiber by reducing the number of generated modes, or attempt to regain lost power through multi-mode heterodyning with delay equalization.

Table 6.4 describes the signal-to-noise ratios that characterize a homodyne detection receiver.

The next class of communications receivers takes advantage of pulse code modulation (PCM), where the digital data streams are superimposed in one way, or another, onto the laser carrier. The simplest example of PCM is intensity modulation of the laser carrier, or "ON-OFF Keying (OOK)", These approaches sometimes take advantage of the

Table 6.4 Signal-to-noise ratios for a homodyne detection receiver

Electrical SNR	$$\mathrm{SNR} \;=\; \frac{4\,G^2\,\left(\frac{\eta q}{hf_c}\right)^2 P_c\,P_o\,R_L}{G^2\,q\,\left[\frac{\eta q}{hf_c}\left(P_o + P_B\right) + I_D\right]B_0\,R_L \;+\; 4\,k\,T\,B_0}$$
Electrical SNR when the local oscillator power is large, and the thermal noise and the shot noise terms derived from the background radiation and dark current are negligible compared to the local oscillator shot noise	$$\mathrm{SNR} \;=\; \frac{4\,\eta\,P_c}{hf_c\,B_0}$$

techniques described in the previous section. Table 6.5 describes the probability of bit error for intensity modulation (IM) / OOK communications, which contains the signal and noise detected photo-electron dependencies. In the upper part of the table, we have

$$\mu_{S,B} = \left(q\eta\tau_B/hf\right) P_S \qquad (6.18)$$

and

$$\mu_{N,B} = \left(q\eta\tau_B/hf\right) P_N \qquad (6.19)$$

being the average number of photo-electrons per bit from the laser signal and average number of photo-electrons per bit from background radiation and dark current, respectively. In addition, $\Gamma(n, x)$ is the incomplete gamma function, given by

$$\Gamma(n, x) \equiv \int_0^x t^{n-1}\, e^t\, dt \qquad (6.20)$$

and τ_B is the bit period. The parameter k_D is the detection threshold and the value that minimizes the probability of error is set by the largest integer, k_T, which obeys the following equation:

$$k_T = \frac{\mu_{S,B}}{1 + \left(\mu_{S,B}/\mu_{N,B}\right)} \qquad (6.21)$$

Figure 6.5 redraws Figure 4.2 (OOK systems) using the Pratt notation; K_s is replaced by $\mu_{S,B}$ and K_b is replaced by $\mu_{N,B}$.

In the lower part of Table 6.5, we have the probability of bit error for thermal noise-limited IM/OOK. For an ideal filter with a band-pass from 0 to $\frac{1}{2}B_0$, the thermal noise variance is given by

$$\sigma_{i_{TH}}{}^2 = \frac{2\,k\,T\,B_0}{R_L} \qquad (6.22)$$

By setting $B_0 = 2/\tau_B$ Eq. (6.22) becomes

$$\sigma_{i_{TH}}{}^2 = \frac{2\,k\,T}{\tau_B\,R_L} \qquad (6.23)$$

The detection threshold is given by

$$i_T = \frac{q}{2\,\tau_B}\,\mu_{S,B} \qquad (6.24)$$

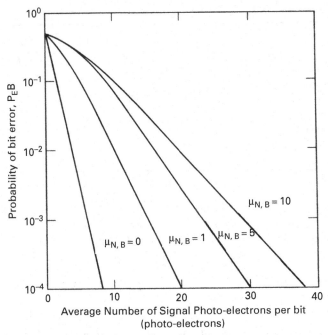

Figure 6.5 Probability of detection error for IM/OOK direct detection optical communications under system shot noise limited conditions

The function erfc(z) is the complementary error function, which is given by

$$\mathrm{erfc}(z) = 1 - \mathrm{erf}(z)$$

$$= \frac{1}{\sqrt{2\pi}} \int_x^{\infty} \exp\left[-\frac{z^2}{2}\right] dz \tag{6.25}$$

There is one variation of the above that should be noted. Pratt stated that if a polarizer precedes the optical detector, the average number of noise photo-electrons is equal to

$$\mu_{N,B} = \eta P_B \tau_B \Big/ 2hf_s + I_D \tau_B \Big/ q \tag{6.26}$$

where P_B is the unpolarized background radiation power. In this case, Eq. (6.26) should be used in the probability of bit error calculation shown in Table 6.5.

The simplest form of binary phase modulation (BPM), or binary phase shift key (BPSK), communications is antipodal signaling. This is when a "1" or "0" bit is selected by whether the received phase determined as either 0 or π during the bit period (or vice versa), respectively. There are several types of BPSK designs that can be used. Table 6.6 gives the probability of bit error for three of the most popular techniques. In this table, we have A_c being the field amplitude of the laser carrier, A_o being the field amplitude of the laser oscillator, N_b being the background noise power, and

$$\sigma_{v_N}^2 = q\,G^2 \left(\frac{\eta q}{hf_c}\right) A_o^2 R_L^2 \left(\frac{1}{2} B_0\right) \tag{6.27}$$

Table 6.5 Probability of bit error equations for IM/OOK communications

IM/OOK shot-noise-limited communications	$P_E^B = \frac{1}{2} \left\{ 1 - \sum_{k=k_D}^{\infty} \frac{\exp\{-\mu_{N,B}\}}{k!} \left[(\mu_{S,B} + \mu_{N,B})^k \exp\{-\mu_{S,B}\} - (\mu_{N,B})^k \right] \right\}$
	$= 0.5 \left[1 + \frac{\Gamma(k_T, \mu_{S,B}) - \Gamma(k_T, \mu_{S,B} + \mu_{N,B})}{\Gamma(k_T, \infty)} \right]$
IM/OOK thermal-noise-limited communication	$P_E^B = 1 \Big/ 2 \left(1 - \mathrm{erf}\left[\frac{q\mu_{S,B}}{2\sqrt{2}\tau_B \sigma_{i_{TH}}} \right] \right)$
	$= 1 \Big/ 2 \left(\mathrm{erfc}\left[\frac{q\mu_{S,B}}{2\sqrt{2}\tau_B \sigma_{i_{TH}}} \right] \right)$

Table 6.6 Probability of bit error for three variations of BPSK

Antipodal signaling binary phase modulation / PSK communications	$P_E^B = 0.5 \left(1 - \mathrm{erf}\left[\sqrt{ \left[\frac{\left(G\left(\frac{\eta q}{hf_c}\right) A_c A_o R_L \right)^2}{2\sigma_{v_N}^2} \right] } \right] \right)$
	$= 0.5 \left(\mathrm{erfc}\left[\sqrt{\frac{\eta A_c^2}{hf_c B_0}} \right] \right)$
	$= 0.5 \left(\mathrm{erfc}\left[\sqrt{\mu_{S,B}} \right] \right)$
Heterodyne BPSK with background noise	$P_E^B = \mathrm{erfc}\left[\frac{\mu_{S,B}}{1 + \left(q\eta/hf_c \right) N_b} \right]$
Heterodyne BFSK with background noise	$P_E^B = 0.5 \exp\left[\frac{1}{8} \left(\frac{\mu_{S,B}}{1 + \left(q\eta/hf_c \right) N_b} \right) \right]$

An alternative to the above techniques is coherent digital optical or coherent binary communications. This approach digitally modulates the amplitude, frequency or phase of the laser signal, using heterodyne or homodyne methods previously. Specifically, one can use subcarrier intensity modulation to send digital data. Binary modulation formats have also been developed for phase-shift keying (PSK) and frequency-shift keying (FSK) under homodyne and heterodyne operations, respectively.

Table 6.7 shows the probability of bit error for binary AM subcarrier IM communications systems. In this table, we have

$$[N]_{sc} = 2qB_{sc}G^2 \left[\left(\frac{q\eta}{hf_s} \right) \left(\frac{P_s}{2} + P_B \right) + I_D \right] R_L \qquad (6.28)$$

Table 6.7 Probability of bit error for AM subcarrier IM systems

AM subcarrier IM systems	$P_E^B = 0.5 \left[\text{erfc} \left(\frac{G\left(\frac{q\eta}{hfs}\right) P_S R_L}{4\sqrt{2\,[N]_{sc}\, R_l}} \right) \right]$
Shot-noise-limited AM subcarrier IM systems with $B_{sc} = B_0$	$P_E^B = 0.5 \left[\text{erfc} \left(\frac{\frac{1}{8}\mu_{S,B}}{\sqrt{\mu_{S,B} + 2\mu_{N,B}}} \right) \right]$
Signal-shot-noise-limited AM subcarrier IM systems	$P_E^B = 0.5 \left[\text{erfc} \left(\frac{1}{8}\sqrt{\mu_{S,B}} \right) \right]$
Thermal-noise-limited AM subcarrier IM systems	$P_E^B = 0.5 \left[\text{erfc} \left(\frac{q\mu_{S,B}}{16} \sqrt{\frac{R_L}{kT\tau_B}} \right) \right]$

Table 6.8 Probability of bit error for IM and FM heterodyne detection systems

IM heterodyne detection systems	$P_E^B = 0.5 \left(1 + Q\left[0, \frac{R_T}{\sigma_\chi}\right] - Q\left[\frac{A}{\sigma_\chi}, \frac{R_T}{\sigma_\chi}\right] \right)$
IM heterodyne detection systems for $B_{IF} = B_0 = 2/\tau_b$ and $R_T \approx \sigma_\chi \sqrt{2 + \frac{A^2}{4\sigma_\chi}}$	$P_E^B \approx 0.5 \left(1 + Q\left[0, \sqrt{2 + \frac{\mu_{S,B}}{4}}\right] - Q\left[\sqrt{\mu_{S,B}}, \sqrt{2 + \frac{\mu_{S,B}}{4}}\right] \right)$
	$\approx 0.5 \exp\left[-\frac{\mu_{S,B}}{8}\right]$ for large signal counts
FM heterodyne detection systems for a strong local oscillator	$P_E^B = 0.5 \left(\exp\left[-\frac{\mu_{S,B}}{4}\right] \right)$

Table 6.8 shows the probability of bit error for IM and frequency modulation (FM) heterodyne detection communications systems. In this table, we have $Q(a, b)$ being the Marcum Q function given by

$$Q(a, b) \equiv \int_b^\infty x \exp\left[-\frac{(a^2 + x^2)}{2} \right] I_0(ax)\, dx \tag{6.29}$$

$$\sigma_\chi^2 = [N]_{IF} R_L \tag{6.30}$$

and R_T is the likelihood ratio test envelope threshold found by solving the following transcendental equation:

$$\exp\left[-\frac{A^2}{2\sigma_\chi^2} \right] = I_0\left(\frac{AR}{\sigma_\chi^2} \right) \tag{6.31}$$

One of the best ways to minimize background radiation like solar illumination corrupting communications system performance is peak power detection, which only measures all incoming optical energy within approximately the pulsed laser signal's time duration, its pulse-width. The most natural way to exploit this energy detection approach for communications is through pulse position modulation (PPM). PPM communications was introduced in section 4.6 in Chapter 4. It conveys information by varying the time of occurrence of a single optical pulse within a data sample time period. The sample period, called a

Table 6.9 Probability of bit error for binary PPM

Binary PPM communications	$P_E^P = \sum\limits_{k_1=0}^{\infty} \sum\limits_{k_2=0}^{\infty} Pos\,(k_1,\,K_S + K_N)\,Pos\,(k_2,\,K_N)\,\gamma_{k_1 k_2}$
Binary PPM communications from Pratt	$P_E^P = Q(\sqrt{2\,m_0},\,\sqrt{2\,m_1}) - 1/2\,\exp\left[-(m_0 + m_1)\right]I_0\,(2\sqrt{m_0\,m_1})$

frame, is subdivided into several time windows, called slots. The pulse repetition rate of the pulsed laser systems multiplied by the bits per pulse equals the data rate of the PPM system.

Table 6.9 shows the probability of bit error for binary pulse PPM. In this table we have

$$K_S = \mu_{s,p}\,\tau_B\,/\,2 \tag{6.32}$$

$$K_N = \mu_{N,B}\,\tau_B\,/\,2 \tag{6.33}$$

$$m_1 = K_S + K_N \tag{6.34}$$

$$m_0 = K_N \tag{6.35}$$

Hubbard developed the following useful bound for the bit error probability:

$$P_E^P = \left[\frac{\sqrt{m_0}}{\sqrt{m_1} - \sqrt{m_0}} + \frac{1}{2}\right]\exp\,\left[-(m_0 + m_1)\right]I_0\,(2\sqrt{m_0\,m_1}) \tag{6.36}$$

[2, 7]. Figure 6.5 is a plot of the probability of bit error using the upper equation of Table 6.9 as a function of K_S various background noise photo-electrons, K_N. This figure redraws Figure 4.3 in Pratt's notation. Figure 4.4 is the probability of bit error for binary PPM using the lower equation of Table 6.9.

Table 6.10 shows the probability of bit error for block code, or M-ary, PPM. In this table, the first row gives the general probability of bit error equation for PPM direct detection communications and the second row gives the probability of bit error when the a priori probability of the symbols is uniformly distributed. The third and fourth rows are the parameters for this equation, shot-noise-limited and thermal-noise-limited, respectively. In this table, the value of k_D that minimizes the probability of detection error for the shot-noise-limited case is the greatest integer value of the likelihood ratio test threshold, k_T, which is defined as

$$k_T = \frac{\mu_{S,P} + \ln\,(M-1)}{\ln\left[1 + \left(M\mu_{S,P}\,\Big/\,\mu_{N,P}\right)\right]} \tag{6.37}$$

The detection threshold is for the thermal noise-limited operations is given by

$$i_T = \frac{q}{2\,\tau_P}\mu_{s,p} + \frac{\sigma_{i_{TH}}^2}{\left(q\,/\,\tau_P\right)\mu_{s,p}}\,\ln\,(M-1) \tag{6.38}$$

Table 6.10 Probability of bit error for M-ary PPM communications

PPM direct detection communications	$P_E^P = \sum_{i=1}^{M} p_i \left[1 - (1-P_N^P)^{i-1}\right] + \sum_{i=1}^{M} p_i \left(1-P_N^P\right)^{i-1}\left(1-P_{SN}^P\right)\left(1-(1-P_N^P)^{M-i}\right)$ $+ (1-P_N^P)^{M-1}(1-P_{SN}^P)\sum_{i=1}^{M} p_i(1-p_i)$
PPM direct detection communications for uniform source distribution where $p_i = 1/M$	$P_E^P = \left(1 - \dfrac{P_{SN}^P}{M P_N^P}\right) + \dfrac{(1-P_N^P)^{M-1}}{M P_N^P}\left(P_{SN}^P - P_N^P\right)$
Parameters for PPM/IM shot-noise-limited direct detection communications	$P_{SN}^p = \sum_{k=k_D}^{\infty} \left[\mu_{S,P} + \left(\mu_{N,P}/M\right)\right]^k \dfrac{\exp\left\{-\left[\mu_{S,P}+\left(\mu_{N,P}/M\right)\right]\right\}}{k!}$ and $P_N^P = \sum_{k=k_D}^{\infty} \left[\mu_{N,P}/M\right]^k \dfrac{\exp\left\{-\left[\mu_{N,P}/M\right]\right\}}{k!}$
Parameters for IM/IM thermal-noise-limited direct detection communications	$P_{SN}^P = \int_{i_T}^{\infty} (2\pi\sigma_{i_{TH}}^2)^{-1/2} \exp\left[\dfrac{-\left[i_f - (q/\tau_P)\mu_{S,P}\right]^2}{2\sigma_{i_{TH}}^2}\right] di_f$ and $P_N^P = \int_{i_T}^{\infty} (2\pi\sigma_{i_{TH}}^2)^{-1/2} \exp\left[\dfrac{-\left[i_f\right]^2}{2\sigma_{i_{TH}}^2}\right] di_f$

6.3   Modern system design considerations

More modern systems employ high-power amplifiers such as an erbium-doped fiber amplifier (EDFA) in both the transmitter and receiver subsystems because of their high electrical-to-optical conversion efficiency, forward error correction (FEC) techniques for potential coding gain and optical receiver sensitivities in setting ultimate system performance. In this section, we will highlight the key optical detectors employed today and discuss the systems implications of using an ERDA in the optical receiver systems. An ERDA is probably the most popular high-power amplifier in use today.

High-data-rate fiber optic and free-space optical communications (FOC/FSOC) systems generally use Avalanche photo-detectors (APDs), p-n junction or PIN photo-detectors. Both are made of semi-conductor material. APDs generate a large number of photo-electrons for small input signals to ensure the receiver will detect the most signal possible. Alternately, the PIN photo-detector does not create multiple photo-electrons when illuminated. The latter is usually the preferred approach because APDs create improved receiver sensitivity at the expense of bandwidth and available dynamic range.

Thermal noise in an optical receiver comes from the fact that a receiver at any temperature above $0°$ Kelvin will generate some sort of current in its circuitry, even when no signal stream is present. Called Johnson noise, the variance of the thermal noise current per unit frequency is given by

$$\sigma_{thermal}{}^2 = \frac{4kTB_e}{R_L} \tag{6.39}$$

System shot noise comes from uncertain arrival time of photo-electrons generated by incident light being absorbed by the photo-detector due to quantum mechanical effects. In particular, the resulting DC photo-current, I_s, will have a variance per unit frequency given by

$$\sigma_{shot}{}^2 = 2qB_eGI_s \tag{6.40}$$

Induced noise (signal-signal beat noise), signal-spontaneous (s-sp) beat noise, and spontaneous-spontaneous (sp-sp) beat noise spectra are three new noise sources one is confronted with when using an ERDA, or any other doped fiber amplifier. The source-induced noise will normally be the dominating noise source, but in some applications the other noise terms also will be of importance, as we will soon see. Becker, Olsson and Simpson did an excellent job in describing these sources in their seminal book on ERDAs [4], and cited the following noise variances from these sources:

$$\sigma_{shot}{}^2 \equiv \text{ Shot Noise}$$

$$= 2qB_e(GI_s + I_{sp}) \tag{6.41}$$

$$\sigma_{s-sp}{}^2 \equiv \text{Signal-Spontaneous (s-sp) beat Noise}$$

$$= 2GI_sI_{sp}\left(\frac{B_e}{B_0}\right) \tag{6.42}$$

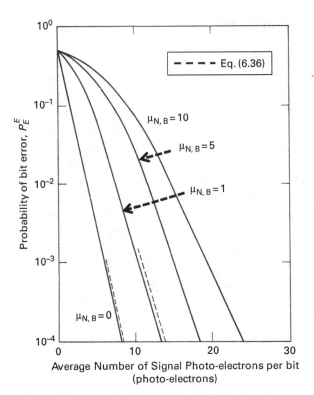

Figure 6.6 Bit error probability for Binary PPM system.

$\sigma_{sp-sp}{}^2 \equiv$ Spontaneous-Spontaneous (s-sp) beat Noise

$$= \frac{1}{2} I_{sp}{}^2 B_e \left((2B_0 - B_e)/B_0{}^2 \right) \tag{6.43}$$

where

$$I_s = \left(\frac{q\eta}{hf} \right) P_s \tag{6.44}$$

$$I_{sp} = 2 q n_{sp} (G-1) B_0 \tag{6.45}$$

$B_e \equiv$ electrical bandwidth
$B_0 \equiv$ optical bandwidth
$\Delta v \equiv$ Laser frequency linewidth
$P_s \equiv$ receiver bit "1" optical power before the amplifier
$n_{sp} \equiv$ inversion parameter [4].
Given all of the above, we define

$$\sigma_{total}{}^2 = \sigma_{shot}{}^2 + \sigma_{thermal}{}^2 + \sigma_{s-sp}{}^2 + \sigma_{sp-sp}{}^2 \tag{6.46}$$

to be the total noise power variance. Figure 6.7 illustrates a typical plot of various amplifier noise powers, their total and the received signal-to-noise ratio as a function of amplifier gain. In this figure, $\lambda = 1.55$ μm, B_e is 7.5 GHz, B_0 is 75 GHz, $n_{sp} = 1.4$, T is 290°K, $\eta = 0.7$, and the received power is −30 dBmW. The electrical signal-to-noise ratio is given by

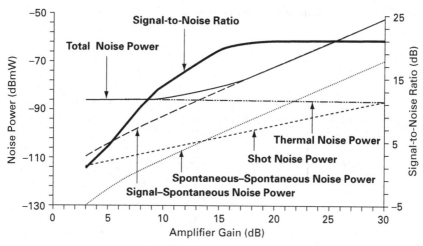

Figure 6.7 Noise powers and signal-to-noise ratio for an erbium-doped fiber pre-amplifier system as a function of amplifier gain.

$$\text{SNR}_e = \left(\frac{G^2 I_s^2}{\sigma_{\text{total}}^2} \right) \tag{6.47}$$

as defined in previous sections. It is clear from this graph that thermal noise typically is the major system noise source when the gain is low. As the gain increases, the in-band beat noise (post detection) from the signal-and-spontaneous (s-sp) and from the spontaneous-and-spontaneous emissions emerges as the dominate noise sources. This is one of the times that signal-induced noise does not dominate. These sources affect the form of the electrical signal-to-noise ratio under high amplifier gain operation.

Example 6.5 For a coherent system with amplitude-shift keying (ASK), the numerator in Eq. (6.47) is replaced by

$$2 G I_s I_{lo} L \tag{6.48}$$

where L represents any loss between the amplifier and the photo-detector, and I_{lo} is the photo-current created by the local oscillator power, P_{lo} [4]. Mathematically, we have

$$I_{lo} = \left(\frac{q \eta}{hf} \right) P_{lo} \tag{6.49}$$

In addition, the shot noise from the local oscillator, $2q I_{lo} B_e$, must be added to the total noise variance given in Eq. (6.46), as well as the local oscillator beat noise, given by

$$\sigma_{lo-sp}^2 = 4 q I_{lo} (G - 1) n_{sp} B_e \tag{6.50}$$

The electrical signal-to-noise ratio given in Eq. (6.13) reduces to

$$SNR_e = \left(\frac{G^2 I_s^2}{\sigma_{s-sp}^2 + \sigma_{sp-sp}^2} \right) \tag{6.51}$$

when $\sigma_{shot}^2 + \sigma_{thermal}^2 \lll \sigma_{s-sp}^2 + \sigma_{sp-sp}^2$ under high amplifier gain levels. This implies that we write

$$SNR_e = \left(\frac{G^2 I_s^2}{\left[2 G I_s I_{sp} \left(B_e / B_0 \right) \right] + 1/2 \, I_{sp}^2 \left(\frac{B_e \, (2 B_0 - B_e)}{B_0^2} \right)} \right) \tag{6.52}$$

$$= \left(\frac{SNR_0^2 \, (B_0 / 2 B_e)}{SNR_0 + 1/2} \right) \tag{6.53}$$

$$= SNR_0 \, (B_0 / 2 B_e) \tag{6.54}$$

as $B_e \ll 2 B_0$ and $SNR_0 \gg \frac{1}{2}$, in general [5]. In the above,

$$SNR_0 = \left(\frac{G I_s}{I_{sp}} \right) \tag{6.55}$$

This shows that electrical signal-to-noise ratio is proportional to the optical signal-to-noise ratio when the non-signal-related noise from the high-power amplifier dominates the receiver system noise. Writing out the optical electrical signal-to-noise ratio, we obtain

$$SNR_0 = \left(\frac{G \, (\eta / hf) \, P_s}{2 \, n_{sp} \, (G-1) \, B_0} \right) \tag{6.56}$$

Substituting Eq. (6.56) into the electrical signal-to-noise ratio equation yields

$$SNR_e = \frac{G \, (\eta / hf) P_s}{2 \, n_{sp} \, (G-1) \, B_0} \, (B_0 / 2 B_e) \tag{6.57}$$

$$= \frac{G \, (\eta / hf) \, P_s}{4 \, n_{sp} \, (G-1) \, B_e} \tag{6.58}$$

$$\approx \frac{\eta \, P_s}{4 \, n_{sp} \, (hf) \, B_e} = \frac{SNR_{\text{Signal Shot Noise}}}{2 \, n_{sp}} \tag{6.59}$$

for $G \gg 1$. Eq. (6.59) shows that electrical signal-to-noise ratio for amplifier noise-limited communications is proportional to the "signal-shot-noise (quantum-limited)" signal-to-noise ratio (B_e replacing B_0), scaled by the inverse of twice the inversion parameter under high amplifier gain conditions. It says that the SNR_e is constant under high gain conditions, which is consistent with Figure 6.5, where SNR_e reaches an asymptote for $G > 16$ dB.

Example 6.6 For $\lambda = 1.55$ μm, $B_0 = 75$ GHz, $B_e = 7.5$ GHz, $n_{sp} = 1.4$, $\eta = 0.7$ and $P_s = -30$ dBmW, Eq. (6.59) equals

$$
\begin{aligned}
\text{SNR}_e &= \frac{\eta P_S}{4 n_{sp} h f B_e} \\
&= \frac{\left((0.7)\left(10^{-6}\,\text{W}\right)\right)}{\left(4\,(1.41)\left(6.34\times10^{34}\,\text{joule-sec}\right)\left(1.93\times10^{14}\,\text{Hz}\right)(7.5\,\text{GHz})\right)} \\
&= 130\,(21.1\,\text{dB})
\end{aligned}
$$

which is the asymptotic gain shown in Figure 6.7 for $G > 16$ dB. Thus, the optical signal-to-noise ratio for this situation is

$$
\begin{aligned}
\text{SNR}_0 &= \text{SNR}_e \left(\frac{2 B_e}{B_0}\right) \\
&= \frac{\left((130)\,(2\times75\,\text{GHz})\right)}{(75\,\text{GHz})} \\
&= 130\,(0.2) = 26\,(14.1\,\text{dB})
\end{aligned}
$$

References

1. A. K. Majumdar and J. C. Ricklin. *Free-Space Laser Communications; Principles and Advances*. Springer, New York (2008).
2. R. M. Gagliardi and S. Karp. *Optical Communications*, 2nd edn. Wiley Series in Telecommunications and Signal Processing, John Wiley and Sons, New York (1995).
3. W. K. Pratt. *Laser Communications Systems*. John Wiley & Sons, New York (1969).
4. P. Becker, N. Olsson and J. Simpson. *Erbium-Doped Fiber Amplifiers Fundamentals and Technology*. Academic Press, New York (1999).
5. E. Sackinger. *Broadband Circuits for Optical Fiber Communication*. Wiley, New York (2005).
6. C. E. Shannon. A mathematical theory of communications. *Bell System Technical Journal*, **27** (1948), pp, 379–423, 623–656.
7. B. Sklar. *Digital Communications: Fundamentals and Applications*, 2nd edn. Prentice Hall, Upper Saddle River, NJ (2001).
8. S. Haykin. *Digital Communications*, 4th edn. John Wiley & Sons (2000).

7 Forward error correction coding

When a digital communications system experiences a very noisy/fading channel, the electrical signal-to-noise ratio may never be strong enough to obtain a low probability of detection error. Shannon proved in 1948 that a coding scheme could exist to provide error-free communications under those conditions. In particular, he showed it was possible to achieve reliable communications, i.e., error-free communications, over an unreliable, i.e., noisy, discrete memory-less channel (DMC), using block codes at a rate less than the channel capacity if the number of letters per code word, and consequently, the number of code words, are made arbitrarily large [1]. This result motivated a lot of research to find the optimum code, i.e., one that minimizes the number of letters and code words, to create reliable communications. The various approaches that have emerged from this research form a class of coding best known as forward error correction (FEC) coding (also called channel coding) [2–4]. Richard Hamming is credited with pioneering this field in the 1940s and invented the first FEC code, the Hamming (7,4) code, in 1950. He, and most others, basically has the originator systematically add generated redundant data to its message. These redundant data allows the receiver to detect and correct a limited number of errors occurring anywhere in the message stream without that person requesting the originator to resend part, or all, of the original message. In other words, the FEC provides a very effective means to correct errors without needing a "reverse channel" to request retransmission of data. This advantage is at the cost of a higher channel data rate if one wants to keep the information data rate the same. These techniques are typically applied in communications systems where retransmissions are relatively costly, or impossible, such as in mobile ad hoc networking when broadcasting to multiple receivers (multicast) [5–7], HF communications and optical communications. For the interested reader, Zhu and Kahn provide a detailed summary of FEC coding applied to the turbulence channel [8, Ch. 7, pp. 303–346].

In this chapter, we will provide the basics on block codes, and highlight the arguably most popular FEC, Reed-Solomon [2–4]. We then will note other important types of FEC schemes that can be employed, based on the authors' bias. This material is not meant to be an exhaustive survey of FEC coding, but rather included in this book to give the reader a basic background and knowledge of FEC to illustrate their utility in optical communications today. Application of FEC to optical systems will be discussed in a subsequent chapter.

7.1   Motivation – real-world binary symmetric channel performance

In the previous chapter, we discussed binary signal modulation schemes for optical channels, which is a form of binary symmetric channel (BSC) communications; that is, where only one of two symbols can be transmitted, usually a "0" and "1". (A non-binary channel would be capable of transmitting more than two symbols.) We noted that the bit transmission is not perfect, and occasionally the receiver gets the wrong bit, leading us to develop equations projecting a bit error rate (BER) for the various schemes as a performance metric. The concept and analysis of BSC communications was originally developed by Shannon and it has proved to be the foundation of communications and information theories. This is because it is one of the simplest channels to analyze and because many problems in communications theory can be quantified by a BSC analysis.

Figure 7.1 shows the layout of a BSC. A transmitter sends a bit, and the receiver receives a bit. It is assumed that the bit is usually transmitted correctly, but that it can be "flipped" with a small probability p, known as the "crossover probability". If X is the transmitted random variable and Y the received variable, then the channel is characterized by the conditional probabilities:

$$\Pr\left(Y = 0 \,|\, X = 0\right) = 1{-}p$$

$$\Pr\left(Y = 0 \,|\, X = 1\right) = p$$

$$\Pr\left(Y = 1 \,|\, X = 0\right) = p$$

$$\Pr\left(Y = 1 \,|\, X = 1\right) = 1{-}p$$

It is assumed that $0 \leq p \leq 1/2$. Detection theory dictates that the optimal form of modulation for a BSC divides the "0"s and "1"s by a threshold that makes the crossover (error) probability equal [9]. There are many realizations of this channel as we have shown in the previous chapter. In the real world other factors bear on the selection of a modulation scheme. For example, if the threshold has to be estimated in real time, an additional error is introduced that must be considered. On the other hand, there are modulation schemes for which the threshold is zero. These approaches offer an advantage. These schemes can be considered as naturally optimum binary symmetric channels. Let us look at this in more detail.

In the last chapter, we showed that for an average power constraint on the communications receiver, OOK signaling is optimum. On the other hand, we know that many

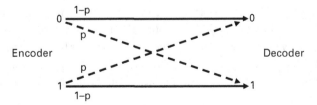

Figure 7.1 Basic block encoder communications system.

lasers have peak power constraints for which the average power used would be one half of the peak power. One could easily double the average power by considering Manchester encoding (2-PPM), which would be constant power. Unfortunately, this approach takes twice the bandwidth, but only yields half the data rate. Another option would be polarization modulation, which solves both problems, but also introduces a measurement of polarization, an equivalent to a zero threshold estimate. This increases system complexity, especially when the effects of the atmosphere are taken into account. For these reasons coherent communications have always been desirable, but are generally used only in short range or space applications. Incoherent techniques find broader utility when the optical scatter and turbulent channels are involved. (We shall discuss this more in Chapters 10 and 11.). The bottom line is that the various optical communications systems will experience large BERs or low link availabilities, negating its potential data rate and spectrum advantages. Since every real system will have some element of dispersion or multipath, this is where the inclusion of FEC coding can help bring desired systems performance back in the BSC. In other words, these systems become more attractive in the high-performance arena through the use of well-established techniques to correct for bit errors and multi-symbol interference. In this chapter, we will introduce some of the coding systems that have evolved over the years which can be applied to real-world optical communication systems.

Example 7.1 One of the key benefits one can obtain from the use of an FEC is improved communications performance by coding the data, at the expense of increased data rate. The parameter most often used to characterize this improvement is coding gain. For a given bit error probability, the coding gain is the reduction in (E_b/N_0) through the use of the code. Mathematically, the coding gain is defined as

$$G_C = \left[\frac{E_b}{N_0} \right]_U (\text{dB}) - \left[\frac{E_b}{N_0} \right]_C (\text{dB})$$

where $(E_b/N_0)_U$ and $(E_b/N_0)_C$ represent the required (E_b/N_0), uncoded and coded.

7.2   Block codes

Block encoders are encoders that take binary input data, group them into segments of L bits, and transmit a resulting code word of

$$N = \left[\frac{\tau_s L}{\tau_c} \right] \text{bits} \tag{7.1}$$

where τ_s is the number of binary bits entering the encoder and τ_c is the number of transmitted channel bits. Figure 7.2 shows the basic system layout. Let $\tilde{a} = (a_{i1}, a_{i2},$

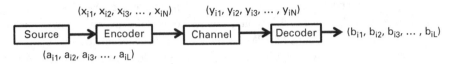

Figure 7.2 Basic Block Encoder Communications System

$a_{i3}, \ldots, a_{iL}$), denote one of the M source input sequences. For each of these $M = 2^L$ input sequences, there is a separate code word provided and we denote the ith code word by

$$\widetilde{x}_i = (x_{i1}, x_{i2}, x_{i3}, \ldots, x_{iN}) \tag{7.2}$$

for $1 \leq i \leq M$.

The rate R_{BC} of a block code is given by

$$R_{BC} = \left[\frac{\log M}{N}\right] = \left[\frac{L \log 2}{N}\right] \tag{7.3}$$

For L/N being an integer and the logarithm base being in base 2, Eq. (7.3) reduces to

$$R_{BC} = \left[\frac{\tau_c}{\tau_s}\right]. \tag{7.4}$$

At the channel output, we receive

$$\widetilde{y}_i = (y_{i1}, y_{i2}, y_{i3}, \ldots, y_{iN})$$

and make our best estimate of which of M inputs was transmitted. The resulting output sequence is denoted by

$$\widetilde{b}_i = (b_{i1}, b_{i2}, b_{i3}, \ldots, b_{iL}).$$

We say we have made a block-decoding error if one, or more, of the L binary bits of the original binary stream, $\tilde{a}$, is in error. As you would expect, there is a delay proportional to $N\tau_c$ in decoding the data, and the complexity of the system goes up as N increases.

7.3   Block coding techniques

In block coding, we group input sequences of length k and encode them into output sequences of $n > k$. (In the last subsection, we had $k = L$ and $n = N$.) Given binary data and all errors assumed to occur independently, we define an (n, k) block code as a set of 2^k code words where $n > k$ and all code words are distinct.

The simplest kind of code words is the so-called parity check codes. A systematic parity check code is one where the first k digits of the n-tuple (i.e., code word) are the information digits that are to be coded, and the last $(n-k)$ digits are linear combinations of the first k digits. The latter digits are called parity check digits. The basic operation in binary coding is modulo-2 addition where if a and b are two symbols, each either a "0" or a "1", then

$$a \oplus b = 0 \qquad \text{if } a = b,$$
$$= 0 \qquad \text{if } a \neq b \qquad (7.5)$$

which have the normal associative, cumulative and distributive properties. We also have the following property:

$$a \oplus b = c, \text{ then } a \oplus (a \oplus b) = b = a \oplus c \qquad (7.6)$$

This last equation implies that in modulo-2, subtraction is equivalent to addition.

The (Hamming) distance between two code words is defined as the number of places in which they disagree. This means that

$$d\,(\tilde{u},\,\tilde{v}\,) \equiv \text{the number of "1"s in the sequence } \tilde{u} \oplus \tilde{v}, \tilde{u} \text{ and } \tilde{v} \text{ being codewords}$$

The Hamming distance has the typical properties of a distance:

$$d\,(\tilde{u},\,\tilde{v}\,) \geq 0, \text{ with equality if and only if } \tilde{u} = \tilde{v}$$

$$d\,(\tilde{u},\tilde{v}) = d\,(\tilde{v},\tilde{u})$$

$$d\,(\tilde{u},\tilde{v}) = d\,(\tilde{u},\tilde{w}) + d\,(\tilde{w},\tilde{v})$$

With the above definitions, we provide the following two theorems.

THEOREM 1. *For a binary symmetric channel, maximum likelihood decoding is equivalent to minimum distance decoding. If the 2^{2k} code words are all equally likely, then this procedure minimizes the probability of error.*

THEOREM 2. *Given code words*

$$\tilde{x}_1, \tilde{x}_2, \tilde{x}_3, \cdots, \tilde{x}_{2k}$$

and a positive number e such that

$$d(\tilde{x}_i, \tilde{x}_j) \geq 2e + 1$$

for all $i \neq j$, then the code can correct all single double, triple, ..., e-tuple errors. Although it will not be shown here, we also find that if

$$d(\tilde{x}_i, \tilde{x}_j) \geq 2e,$$

then all error patterns up to $(e-1)$ errors can be corrected and any e error patterns can be detected. The converse of both statements also is true.

Example 7.2a Let

$$\tilde{x}_1 = (000) \quad and \quad \tilde{x}_2 = (111)$$

Then by the theorem, the minimum distance is equal to

$$d_{min} = 3 = 2e + 1,$$

which implies $e = 1$ and all single errors are correctable if detected. That is, if the estimated received data sequence is (010), then we decode (000).

Example 7.2b Let

$$\tilde{x}_1 = (000) \quad and \quad \tilde{x}_2 = (011)$$

then the minimum distance is given by

$$d_{min} = 2 = 2e,$$

and $e = 1$. In this case, only single errors are detectable, but no errors are correctable. For example, if the estimated received data sequence is (010), we are not able to decide whether the error occurred from (000) becoming (010) or (011) becoming (010).

Example 7.2c Let

$$\tilde{x}_1 = (000) \quad and \quad \tilde{x}_2 = (001),$$

then the minimum distance is given by

$$d_{min} = 1$$

and $e = 0$. In this case, no set number of correctable or detectable errors is guaranteed.

While the above theorem provides us with a way to check on the error-correcting capability of codes, this is very time-consuming when dealing with a very large number of code words. For example, a (23,12)-block code has 2^{23} different possible code words from which to choose the 2^{12} input sequences from. To also decode such code, a "code book" containing 2^{23} entries will be required since each n-tuple that can be received must be mapped to the code word to which it is closest. To remedy this situation, results from group theory are necessary. We will see that shortly.

Let u_i represent the ith input digit to the encoder. We define two possible mathematical operations:

$$x_i = u_i \quad \text{for } 1 \leq i \leq k \tag{7.7}$$

and

$$x_i = \sum_{j=1}^{k} g_{ji} u_i \quad \text{for } k+1 \leq i \leq n \tag{7.8}$$

where

$$g_{ji} = [\text{``0'' or ``1''}] \tag{7.9}$$

or

$$x_i = \sum_{j=1}^{k} g_{ji} u_i \ \forall i \tag{7.10}$$

with

$$g_{ji} = 1 \ \text{for} \ j = i \ \ 1 \leq i \leq k \tag{7.11a}$$

$$= 0 \ \text{for} \ j \neq i \ \ 1 \leq i \leq k \tag{7.11b}$$

The above mathematical operation can be written in matrix formulation. Define

$$\tilde{G} = \begin{bmatrix} g_{11} & g_{12} & \cdots & g_{1n} \\ g_{21} & g_{22} & \cdots & g_{2n} \\ \vdots & \vdots & \ddots & \vdots \\ g_{k1} & g_{k2} & \cdots & g_{kn} \end{bmatrix} \tag{7.12}$$

to be a $k \times n$ matrix called the generator matrix. This name follows because it "generates" a y code word in the following manner

$$\tilde{x} = \tilde{u} \, \tilde{G} \tag{7.13}$$

with

$$\tilde{x} \equiv 1 \times n \text{ row vector}$$

and

$$\tilde{u} \equiv 1 \times k \text{ row vector}$$

For a systematic parity check code, the generator matrix looks like

$$\tilde{G} = \begin{bmatrix} 1 & 0 & \cdots & 0 & g_{1,k+1} & g_{1,k+2} & \cdots & g_{1n} \\ 0 & 0 & \cdots & 0 & g_{2,k+1} & g_{2,k+2} & \cdots & g_{2n} \\ \vdots & \vdots & \ddots & \vdots & \vdots & \vdots & \ddots & \vdots \\ 0 & 0 & \cdots & 1 & g_{k,k+1} & g_{k,k+1} & \cdots & g_{kn} \end{bmatrix} \tag{7.14}$$

or

$$\tilde{G} = \left[\tilde{I} \mid \tilde{P} \right] \tag{7.15}$$

where

$$\tilde{I} \equiv k \times k \text{ identity matrix} \tag{7.16}$$

and

$$\tilde{P} \equiv k \times (n-k) \text{ matrix of parity check bits} \tag{7.17}$$

The transpose of the generator matrix is called the parity check matrix and is defined to be

$$\tilde{H} = \begin{bmatrix} \tilde{P} \\ \tilde{I} \end{bmatrix} \tag{7.18}$$

Note that since

$$x_i = \sum_{j=1}^{k} g_{ji} x_i \quad \text{for } k+1 \leq i \leq n \tag{7.19}$$

as $x_i = u_i$ for $1 \leq i \leq k$, then

$$x_i \oplus x_i = 0 = \left[\sum_{j=1}^{k} g_{ji} x_i \right] \oplus x_i \tag{7.20}$$

Since Eq. (7.24) is true for any $k+1 \leq i \leq n$, we have

$$\tilde{x} \, \tilde{H} = \tilde{0} \tag{7.21}$$

Conversely, any arbitrary sequence that satisfies Eq. (7.25) also satisfies Eq. (7.23). The implication of the above is that the sequence is a code word.

Consider now any received sequence

$$\tilde{w}_i = (w_{i1}, w_{i2}, w_{i3}, \ldots, w_{iN})$$

We define the syndrome as a $1 \times (n-k)$ row vector by the equation

$$\tilde{S} = \tilde{w} \, H \tag{7.22}$$

with the ith component of the syndrome given by

$$S_i = \left[\sum_{j=1}^{k} g_{ji} w_i \right] \oplus w_i \quad \text{for } 1 \leq i \leq n-k \tag{7.23}$$

This implies that $S_i = 1$ if and only if R_{k+1}, the ith check digit, differs from ith check digit as computed from the information digit.

Specifically, we define the error sequence by the relations

$$\tilde{R} = \tilde{x}_m \oplus \tilde{e}_m \tag{7.24}$$

and

$$\tilde{S} = \tilde{R} \tilde{H} = (\tilde{x}_m \oplus \tilde{e}_m) \tilde{H} = \tilde{x}_m \tilde{H} \oplus e_m \tilde{H} = \tilde{e}_m \tilde{H} \tag{7.25}$$

Two observations can be made about Eq. (7.25):

(1) There are two other ways for the syndrome to equal the null vector besides the received sequence equalling the input sequence. One way is that the error sequence is identically equal to a code word, or the modulo-2 convolution of the input and error sequences yield one of the code words.

(2) In general, there is more than one error sequence that will satisfy this equation for the particular input sequence transmitted. The particular one to choose to decode with depends upon the channel. For example, for a binary symmetric channel, we know

we decode that code word whose Hamming distance is closest to the received sequence. Specifically, we calculate the syndrome and then find that error sequence that has the smallest weight that satisfies

$$\tilde{S} = \tilde{e}_m \tilde{H} \qquad (7.26)$$

To further reduce the above procedure, we must now employ the group theory mentioned above.

We will begin with the definition of group. A group is a set G together with an operation (say *) with the following properties:

1. if $a \in G$ and $b \in G$, then $a * b \in G$ [*Closure*]
2. if $a * (b*c) = (a*b)*c$, then $a,b,c \in G$ [Associativity]
3. There exists an element $\in G$ such that $a * e = a = e * a$ for all $a \in G$ [Identity]
4. For each $a \in G$, there exists an $a^{-1} \in G$ such that $a^{-1}*a = e = a * a^{-1}$ [Inverse]

In addition, if $a * b = b * a$ for each $a \cdot b \in G$, then the group is called abelian, or a commutative, group.

Example 7.3 The following are example groups:

1. integers under ordinary addition
2. binary sequences under modulo-2 addition.

A subgroup F of G is a set F of G that is a subset of G and which forms a group itself using the group operation * of G.

Example 7.4 For a group, choose all integers under ordinary addition. The group all even integers under ordinary addition qualify as a subgroup of the initial group of integers.

Consider an array of elements formed in the following manner.

$$\begin{bmatrix} f_1 = e & f_2 & \cdots & f_n \\ g_1 & g_1 * f_2 & \cdots & g_1 * f_n \\ \vdots & \vdots & \ddots & \vdots \\ g_m & g_m * f_2 & \cdots & g_m * f_n \end{bmatrix} \qquad (7.27)$$

The first row in the matrix above has as elements the numbers of the subgroup F, with the identity element in the leftmost position, and with the elements of F used once and only once. The second row is formed by taking any element of g_1 not in F and forming the element $g_1 * f_i$. The process is continued by choosing the leftmost element of each row, some member of G that has not previously appeared within the array, and operating it on the first row elements f_i until all elements of G have been used. Each row of the array is called a coset and the leftmost elements are called coset leaders. The following theorem results:

THEOREM 3.   *Every element of* G *is in one and only one coset of a subgroup* F.

Example 7.5 Consider the group of all integers under ordinary addition, i.e., . . ., −2, −1, 0, 1, 2. . . . .. Let F be a subgroup of all multiples of the integer 4. The resulting array is

$$
\begin{bmatrix}
0 & 4 & -4 & 8 & -8 & 12 & -12 & \cdots \\
1 & 5 & -3 & 9 & -7 & 13 & -11 & \cdots \\
2 & 6 & -2 & 10 & -6 & 14 & -10 & \cdots \\
3 & 7 & -1 & 11 & -5 & 15 & -9 & \cdots
\end{bmatrix}
$$

The topmost row is the subgroup F and leftmost column contain the coleaders. The rows below the topmost row are the cosets.

Note in the above array that no element is duplicated. Also, if say 5 was chosen as the coset leader instead of 1, the above array would have been written as

$$
\begin{bmatrix}
0 & 4 & -4 & 8 & -8 & 12 & -12 & \cdots \\
5 & 9 & 1 & 13 & -3 & 17 & -7 & \cdots \\
6 & 10 & 2 & 14 & -2 & 18 & -6 & \cdots \\
7 & 11 & 3 & 15 & -1 & 19 & -5 & \cdots
\end{bmatrix}
$$

The coset elements are identical to the above respective cosets, but in a different order of presentation. This suggests group theory applies.

The connection of group theory to coding can be seen in the following theorems.

THEOREM 4.   *If*

$$\{\tilde{x}\} = \tilde{x}_1, \tilde{x}_2, \tilde{x}_3, \ldots, \tilde{x}_{2^k}$$

represents the set of code words in a parity check code, then it is a group under modulo-2 addition.

THEOREM 5.   *Given a group code with elements*

$$\{\tilde{x}\} = \tilde{x}_1, \tilde{x}_2, \tilde{x}_3, \ldots, \tilde{x}_{2^k},$$

then one can correct up to e *errors if the minimum weight of any code word, excluding the all zero code word, is* 2e + 1.

If we consider the set of all binary n-tuples, this set forms a group under modulo-2 addition. If we consider a group code as a subgroup of this group of n-tuples, we can form a coset array.

THEOREM 6.   *Two vectors* Ã₁ *and* Ã₂ *are in the same coset if and only if their syndromes are equal.*

Given that we have received a vector $\tilde{A}$ resulting in a syndrome corresponding to the jth coset. These are 2k error patterns that could have placed a code word into the jth coset, each corresponding to one of the 2^k code words. Since the probability of the occurrence of an error sequence increases as the number of "1"s decreases, it follows that minimum distance decoding is equivalent to selecting that error pattern out of the 2^k available patterns that has the minimum weight. In other words, if the coset array is chosen such that each coset leader has a minimum weight among all possible sequences in its coset, the decoding procedure is as follows:

1. Calculate the syndrome

$$\tilde{S} = \tilde{A}\,\tilde{H}$$

   This uniquely determines which coset $\tilde{A}$ is in.
2. Form the sum of $\tilde{A}$ and the coset leader of the syndrome. This gives us the decoded code word.

Note that this last step is equivalent to decoding $\tilde{A}$ as the head of the column in which $\tilde{A}$ appears. The importance of the above is that we now can decode by storing only the $2n-k$ syndromes and their respective coset leaders, rather than all $2n$ possible binary n-tuples.

Example 7.6 Consider the following generator matrix of a (7,4) code, whose rate is

$$R_{BC} = \left[\frac{\log M}{N}\right] = \left[\frac{k\log 2}{n}\right] = 4/7$$

We write

$$\tilde{G} = \begin{bmatrix} 1 & 0 & 0 & 0 & 1 & 1 & 1 \\ 0 & 1 & 0 & 0 & 0 & 1 & 1 \\ 0 & 0 & 1 & 0 & 1 & 1 & 0 \\ 0 & 0 & 0 & 1 & 1 & 0 & 1 \end{bmatrix} = \begin{bmatrix} \tilde{I} & | & \tilde{P} \end{bmatrix}$$

where

$$\tilde{P} = \begin{bmatrix} 1 & 1 & 1 \\ 0 & 1 & 1 \\ 1 & 1 & 0 \\ 1 & 0 & 1 \end{bmatrix}$$

Notice that we have a $k = 4$ by $n = 7$ matrix for the generator matrix. This implies

$$\tilde{H} = \begin{bmatrix} \tilde{P} \\ \hline \tilde{I} \end{bmatrix} = \begin{bmatrix} 1 & 1 & 1 \\ 0 & 1 & 1 \\ 1 & 1 & 0 \\ 1 & 0 & 1 \\ 1 & 0 & 0 \\ 0 & 1 & 0 \\ 0 & 0 & 1 \end{bmatrix}$$

This results in Table 7.1 for code words $\tilde{a}$.

Table 7.1 Block coding of input sequence $\tilde{a}$

No.	$\tilde{a}$	$\tilde{x} = \tilde{a}\,\tilde{G}$
1	0000	0000000
2	1000	1000111
3	1100	1100100
4	1110	1110010
5	1111	1111111
6	0100	0100011
7	0110	0110101
8	0111	0111000
9	0010	0010110
10	0011	0011011
11	0001	0001101
12	1010	1010001
13	1011	1011100
14	1001	1001010
15	1101	1101001
16	0101	0101110

Assume the code word

$$\tilde{x} = (1100100)$$

is transmitted and

$$\tilde{R} = \tilde{x}_m \oplus e_m = (1110100)$$

is received. Then,

$$\tilde{S} = \tilde{R}\tilde{H} = [1110100] \begin{bmatrix} 1 & 1 & 1 \\ 0 & 1 & 1 \\ 1 & 1 & 0 \\ 1 & 0 & 1 \\ 1 & 0 & 0 \\ 0 & 1 & 0 \\ 0 & 0 & 1 \end{bmatrix} = [110]$$

It is clear from this matrix multiplication that an error has been made since the syndrome is not the null vector. To decode, we must determine the coset leader corresponding to the above row vector. We now must form the coset array using the above 16 codes words, taking as the coset leader those 7-tuples with minimum weight. In our case, $W_{min} = 3$ and the array is given in Table 7.2. Corresponding to each coset leader, we now compute the $2^{n-k} = 2^{7-4} = 2^3 = 8$ syndromes, which is shown in Table 7.3.

Since the syndrome of the received sequence was (110), the corresponding error pattern is (0010000) if only one error has occurred. This is not to say that this has to be

Table 7.2 Coset array for (7,4) code

Code words	0000000	1000111	1100100	1110010	1111111	0100011	0110101	0111000	0010110	0011011	0001101	1010001	1011100	1001010	1101001	0101110
Correctable error patterns	1000000	0000111	0100100	0110010	0111111	1100011	1110101	1111000	1010110	1011011	1001101	0010001	0011100	0001010	0101001	1101110
	0100000	1100111	1000100	1010010	1011111	0000011	0010101	0011000	0110110	0111011	0101101	1110001	1111100	1101010	1001001	0001110
	0010000	1010111	1110100	1100010	1101111	0110011	0100101	0101000	0000110	0001011	0011101	1000001	1001100	1011010	1111001	0111110
	0001000	1001111	1101100	1111010	1110111	0101011	0111101	0110000	0011110	0010011	0000101	1011001	1010100	1000010	1100001	0100110
	0000100	1000011	1100000	1110110	1111011	0100111	0110001	0111100	0010010	0011111	0001001	1010101	1011000	1001110	1101101	0101010
	0000010	1000101	1100110	1110000	1111101	0100001	0110111	0111010	0010100	0011001	0001111	1010011	1011110	1001000	1101011	0101100
	0000001	1000110	1100101	1110011	1111110	0100010	0110100	0111001	0010111	0011010	0001100	1010000	1011101	1001011	1101000	0101111

Table 7.3 Table of syndromes for the coset leaders

Coset leader	Syndromes
0000000	000
1000000	111
0100000	011
0010000	110
0001000	101
0000100	100
0000010	010
0000001	001

the error sequence that really occurred; rather this is the most likely one if only one error has occurred. Thus, we have

$$\tilde{R} \oplus \tilde{e} = (1110100) \oplus (0010000)$$
$$= (1100100)$$

which turns out to be the code word that was actually sent. Note that this code word appears at the top of the column of the coset array in which the receiver sequence appears.

On the other hand, we could assume the received sequence was

$$\tilde{R} = (1110100) \oplus (1010000)$$
$$= (0110100)$$

The syndrome is (001) and the corresponding coset leader would be (0000001) using our procedure. This would decode to

$$(0000001) \oplus (0110100) = (0110100)$$

which differs from the transmitted code word in three of the seven places.

To determine the probability of a word decoding error for a group code, we note that the correct decoding will take place if and only if the error pattern is the coset leader. That is, if N_i is the number of coset leaders of weight I, we have

$$P_c \equiv \text{probability of a correctly decoded word}$$

$$= \sum_{i=0}^{n} N_i \varepsilon^i (1-\varepsilon)^{n-i} \qquad (7.28)$$

and

$$P_E \equiv \text{probability of a decoded word error}$$
$$= 1 - P_c \qquad (7.29)$$

7.4   Reed-Solomon coding

One of the most popular FEC methods is Reed-Solomon coding. Its uses are many. For example, AOL used Reed-Solomon (R-S) coding to compress attachments in their e-mail systems, reducing the needed bandwidth when the firm was just getting started [10]. It is also used in CD players to allow high-quality audio generation from cheap electronic components. Juarez *et al.* have even used R-S coding in helping mitigate channel fading in optical communications by correcting errors after running the received 10 giga-bits per second (Gbps) laser beam through Adaptive Optics and their Optical Automatic Gain Control [5,11]. This subsection highlights the basics and characteristics of R-S coding following Jonathon Y. C. Cheah [4, pp. 267–294] and Bernard Sklar [2, Ch. 8]; the reader is encouraged to look at these references, as well as references [12–14] for more details, implementations and examples.

Fundamentally, R-S codes are systematic linear block codes, residing in a subset of the BCH codes called non-binary BCH codes.[1] They are a block because the original message is split into fixed-length blocks and each block is split into m-bit symbols; linear because each m-bit symbol is a valid symbol; and systematic because the transmitted information contains the original data with additional parity bits appended. They are built through the use of finite-field arithmetic using a Galois Field of n elements, denoted by GF(n) [4, pp. 267–294].[2] Each code word has n coordinates and its code is referenced to have length n. For commercial cellular digital packet data, $n = 2^6 = 64$ is the number of elements in the field and each code symbol has six bits [ibid.]. Although each code word can be created using k information symbols that can take on all possible values, the use of the Galois field means that the information symbols can only be selected from that field and only can each take on n different values. They are specified as RS (n,k) codes, with m-bit symbols specified by length n and dimension k. This means that the encoder takes k data symbols of m bits each, appends ($n - k$) parity symbols, and produces a code word of n symbols (each of m bits). Figure 7.3 illustrates an RS (63,47) code word.[3]

Code Word (63 bytes)

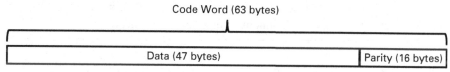

Figure 7.3 Layout of a RS (63, 47) Code Word used in Cellular Communications, where the Number of Parity Bits equals 16 bits[(63-47) bits].   (In this figure, we use 6-bit bytes rather than 8-bit bytes.)

[1] BCH codes form a class of parameterized error-correcting codes that were invented by Hocquenghem, and independently in 1960 by Bose and Ray-Chaudhuri; the acronym BCH comprises the initials of these inventor's names in alphabetical order. The principal advantage of BCH is that they can be decoded using syndrome decoding, which we discussed in the last subsection.

[2] R-S codes are based on a specialized area of mathematics known as Galois fields (aka finite fields). Here, these codes make use of Galois fields of the form GF (n) = GF (2^m), where the n elements of th field can be represented by m binary bits.

[3] Cheah notes that cellular communications use a 6-bit byte rather than an 8-bit byte, the standard convention in electrical engineering, and pose no more problem than some shift and rotation operation in computer programs [4, p. 267].

The R-S codes used today are mostly non-binary cyclic code and as a result, the code length is $(n-1)$ rather than n [ibid.]. Continuing the cellular digital packet data example, they use RS (63, 47) codes, highlighted in Figure 7.3. A cyclic code is where for any code word $\{m_0, m_1, m_2, \ldots m_{n-1}\}$, the cyclically shifted word $\{m_1, m_2, \ldots, m_{n-1}, m_0\}$ is also a code word; the advantage of a cyclic code is that it can always be created by a generator polynomial [ibid.]. Cheah provided a general notation for an RS (n, k) code polynomial in the form

$$p(x)p_0 + p_1x + p_2x^2 + \ldots + p_{n-k-1}X^{n-k-1} \tag{7.30}$$

This definition implies that each code word is interpreted as code polynomial, and any resulting code polynomial is a multiple of $p(x)$. This is a convenient means for mapping symbols into code words, thereby making hardware very easy to design and fabricate for FEC purposes, the reason for its popular use today. This polynomial approach can be applied to any initial data set, so one can write the coding approach as

$$c(x) = d(x)g(x) \tag{7.31}$$

where $c(x)$ is the code word polynomial and $d(x)$ is the information data polynomial. The dimension of a code with a degree $2q$ generator polynomial is $k = n-2q-1$, and the code polynomials are valid up to $n-2$. The error correction capability is $(n-k)/2$ symbols.

Example 7.7 Sklar gives an example of why non-binary codes such as RS (n,k) codes have an advantage over binary codes [2]. Consider a binary code $(n, k) = (7,3)$ code. The entire n-tuple space contains $2^n = 2^7 = 128$ n-tuples, of which $2^k = 2^3 = 8$ are code words. This latter value is 1/16 of the former. For a non-binary $(n, k) = (7,3)$ code, each symbol is composed of 3 bits and the n-tuple space amounts to

$$2^{nm} = 2^{21} = 2\,097\,152, n\text{-tuples.}$$

The number of code words is $2^{km} = 2^9 = 512$. In this case, only 1/4096 of the n-tuples are code words. In other words, when dealing with non-binary symbols, each is made of up of m bits, and there is only a small fraction of the possible code words that need be considered as code words. That fraction decreases for increasing m. The importance of this is that a large code minimum distance can be created.

Sklar states that the RS (n,k) codes work well to correct burst errors; that is, they are effective for channels that have memory [2]. He also says they can be used efficiently on channels where the set of input symbols is large. An interesting feature of the R-S code is that as many as two information symbols can be added to an R-S code of length n without reducing its minimum distance. This extended RS (n,k) code has length $n+2$ and the same number of parity check symbols as the original code. Odenwalder showed that the R-S decoded symbol-error probability, P_{SE}, can be written in terms of the channel symbol-error probability, p_c [13]; specifically, we can write

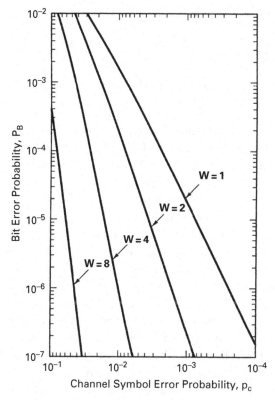

Figure 7.4 P_B versus p_C for 32-ary orthogonal signaling and n = 31, w-error correcting Reed-Solomon coding [16].

$$P_{SE} = \frac{1}{2^m-1} \sum_{j=2w+1}^{2^m-1} j \left[\frac{(2^m-1)!}{j!(2^m-1-j)!} p_c^j (1-p_c)^{2^m-1-j} \right] \tag{7.32}$$

where

$$w = \left| \frac{d_{\min}-1}{2} \right| = \left| \frac{n-k}{2} \right| \tag{7.33}$$

Sklar also stated that the bit-error probability can be upper bounded by the symbol-error probability for specific modulation types [2]. For M-ary FSK modulation with $M = 2^m$, the relationship between the bit error probability P_B and P_{SE} is given as:

$$\frac{P_B}{P_{SE}} = \frac{2^{m-1}}{2^m-1} \tag{7.34}$$

Figure 7.4 shows P_B versus the channel symbol-error probability p_c, using eqns (7.3) and (7.4) for various w-error-correcting, 32-ary orthogonal Reed-Solomon codes with $n = 31$ (thirty-one 5-bit symbols per code block) [13].

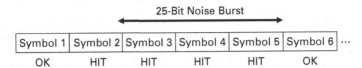

Figure 7.5 Illustration of a data block disturbed by a 25-dB noise burst.

Example 7.8 Sklar gives the following example to show why R-S codes perform well against burst noise. [6]. Consider the R-S code, RS (255,247), where each symbol is made up of $m = 8$ bits. Since $n-k = 8$, Eq. (7.33) implies that this code can correct any four symbol errors in any block of 255 bits. Assume we have a noise burst occurring at some place in symbol 2 and lasting 25 bits in duration. Figure 7.5 illustrates this situation. In this figure, we see that the noise burst affects four symbols only since $m = 8$ bits, i.e., 25 bits is greater than total bits for 3 symbols. Since $t = 4$, the RS (255,247) decoder will correct any four-symbol errors without regard to the type of damage suffered by the symbol. In other words, when a decoder corrects a symbol, it replaces the incorrect symbol with the correct one, whether the error was caused by one bit being corrupted or all eight bits being corrupted. So, if a symbol is wrong, it might as well be wrong in all of its bit positions. This gives an R-S code a tremendous burst-noise advantage over binary codes, even allowing for the interleaving of binary codes. It also should be clear from this example that if the 25-bit noise disturbance had occurred randomly in the data block rather than as a contiguous burst across one segment, many more than four symbols would be affected (as many as 25 symbols might be disturbed). Of course, that would be beyond the capability of the RS (255,247) code.

For a FEC to mitigate the effects of noise, the noise duration has to represent a relatively small percentage of the code word. To ensure that this happens most of the time, Sklar points out that the received noise should be averaged over a long period of time, reducing the effect of a "freak streak of bad luck". In particular, he states that error-correcting Reed-Solomon codes become more efficient (error performance improves) as the code block size increases; this makes this type of code an attractive choice whenever long block lengths are desired [14]. One can see this clearly in Figure 7.6, where the rate of the code is held at a constant 7/8, while its block size increases from $n = 32$ symbols (with $m = 5$ bits per symbol) to $n = 256$ symbols (with $m = 8$ bits per symbol). Thus, the block size increases from 160 bits to 2048 bits.

7.5   Other important FEC schemes

As noted in the introduction, there are many approaches to error correction in order to find the optimum approach. Sklar's seminal book on digital communications covers most of them and the interested reader is directed to that text to get a more detailed look at all

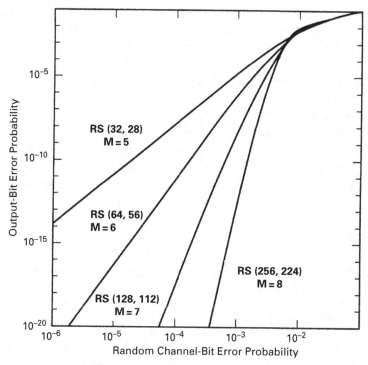

Figure 7.6 Reed-Solomon rate 7/8 decoder performance as a function of symbol size.

the FEC approaches as a starting point. We will summarize these approaches in no particular order.

Elias was the first to investigate convolutional codes in 1955. It basically is the convolution of the input data sequence with a filter impulse response, paralleling what a digital filter does. Improvements in this approach were developed by others; e.g., Wozencraft, Fano and Massey introduced sequential convolutional coding in 1963. The most notable development was by Viterbi in 1967, who achieved optimum error correction performance through the use of a maximum-likelihood algorithm. Viterbi decoding allows asymptotically optimal decoding efficiency with increasing constraint length of the convolutional code, but at the expense of exponentially increasing complexity. His scheme had a shorter decoding delay than other existing sequential decoding schemes.

Low-density parity-check (LDPC) codes were first introduced by Robert G. Gallager in his PhD thesis in 1960, but did not find much practical use at the time because of the computational complexity in creating the encoder and decoder, and the introduction of Reed-Solomon codes. LDPC codes have recently been re-discovered as an efficient linear block code. They have been shown to perform very close to the Shannon channel capacity (the theoretical maximum) k length. Practical implementations can draw heavily from the use of parallelism.

Turbo codes emerged in 1993 as the first practical codes to closely approach the channel capacity, a theoretical maximum for the code rate at which reliable communication is still possible given a specific noise level. The first class of turbo code was the parallel concatenated convolutional code (PCCC). Since then, many other classes of turbo code have been discovered, including serial versions and repeat-accumulate codes. Iterative turbo decoding methods have also been applied to more conventional FEC systems, including Reed-Solomon corrected convolutional codes. Turbo codes are finding use in (deep space) satellite communications and other applications where designers seek to achieve reliable information transfer over bandwidth- or latency-constrained communication links in the presence of data-corrupting noise [15]. Turbo codes are nowadays competing with LDPC codes, which provide similar performance.

References

1. C. E. Shannon. A mathematical theory of communication. *Bell Syst. Technol. J.*, **27**, pp 379–423 & pp 623–656, 1948
2. B. Sklar. *Digital Communications: Fundamentals and Applications*, 2nd edn. Prentice Hall, Upper Saddle River, NJ (2001).
3. S. Haykin. *Digital Communications*, 4th edn. John Wiley & Sons (2000).
4. J. Y. C Cheah. *Practical Wireless Data Modem Design*. Artech House, Boston, MA (1999).
5. L. B. Stotts, J. Foshee, B. Stadler, *et al.* Hybrid optical RF communications. *Proceedings of the IEEE*, **97**(6) (2009), pp. 1109–1127.
6. L. B. Stotts, S. Seidel, T. Krout and P. Kolodzy. MANET gateways: interoperability via the network, not the radio. *IEEE Communications Magazine*, (June 2008), pp. 2–10.
7. S. Siedel, T. Krout and L. B. Stotts. An adaptive broadband mobile ad-hoc radio backbone system: DARPA NetCentric demonstration – Ft. Benning, GA, January 2006 (invited paper). Third Annual IEEE Communications Society Conference on Sensor, Mesh, and Ad Hoc Communications and Networks (SECON), IEEE Workshop on Networking Technologies for Software Defined Radio (SDR) Networks, Reston, VA, September 25, 2006.
8. A. K. Majumdar and J. C. Ricklin. *Free-Space Laser Communications; Principles and Advances*. Springer, New York (2008).
9. R. M. Gagliardi. *Introduction to Communications Engineering*, 2nd edn. John Wiley & Sons, New York (1988), pp. 292–293.
10. I. S. Reed, private communication.
11. J. C. Juarez, D. W. Young, J. E. Sluz and L. B. Stotts. High-sensitivity DPSK receiver for high-bandwidth free-space optical communication links. *Optics Express*, **19**(11) (2011), pp. 10789–10796.
12. I. S. Reed and X. Chen. *Error-Control Coding for Data Networks*. Kluwer Academic, Boston, MA (1999).
13. J. P. Odenwalder. *Error Control Coding Handbook*. Linkabit Corporation, San Diego, CA (1976).
14. I. S. Reed and G. Solomon. Polynomial codes over certain finite fields. *SIAM Journal of Applied Mathematics*, **8** (1960), pp. 300–304.
15. D. Divsalar and F. Pollara. Turbo codes for deep-space communications. *TDA Progress Report*, 42–120 (February 15, 1995).

8 Modern communications designs for FOC/FSOC applications

8.1 Introduction

This chapter discusses some of the key aspects of the signal modulation and coding schemes used in FOC and FSOC systems today. Most notably, we will review the use of return-to-zero (RZ) and non-return-to-zero (NRZ) in coding the information streams and see their effect on systems performance, as well as receiver sensitivity.

8.2 Modern signal modulation schemes

Let us begin with some definitions.

Return-to-zero (RZ)

RZ describes a signal modulation technique where the signal drops (returns) to zero between each incoming pulse. The signal is said to be "self-clocking". This means that a separate clock signal does not need to be sent alongside the information signal to synchronize the data stream. The penalty is the system uses twice the bandwidth to achieve the same data-rate as compared to non-return-to-zero format (see next definition).

Although any RZ scheme contains a provision for synchronization, it still has a DC component, resulting in "baseline wander" during long strings of "0" or "1" bits, just like the line code non-return-to-zero. This wander is also known as a "DC droop", resulting from the AC coupling of such signals

---

Example 8.1 Figure 8.1(a) is an example of a binary signal that is encoded using rectangular pulse amplitude modulation with polar return-to-zero code. The "zero" between each bit is a neutral or rest condition, such as a zero amplitude in pulse amplitude modulation (PAM), zero phase shift in phase-shift keying (PSK), or mid-frequency in frequency-shift keying (FSK). That "zero" condition is typically halfway between the significant condition representing a "1" bit and the other significant condition representing a "0" bit.

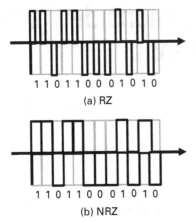

1 1 0 1 1 0 0 0 1 0 1 0

(a) RZ

1 1 0 1 1 0 0 0 1 0 1 0

(b) NRZ

Figure 8.1 The binary signal is encoded using rectangular pulse amplitude modulation with (a) polar return-to-zero code and (b) polar return-to-zero code.

Non-return-to-zero (NRZ)

NRZ describes a signal modulation technique where "1"s are represented by one significant condition (usually a positive voltage) and "0"s are represented by some other significant condition (usually a negative voltage), with no other neutral or rest condition. Figure 8.1(b) illustrates the NRZ version of 8.1(a). The pulses have more energy than an RZ code. Unlike RZ, NRZ does not have a rest state. In addition, NRZ is not inherently a self-synchronizing code, so some additional synchronization technique (for example a run-length limited constraint or a parallel synchronization signal) must be used to avoid bit slip. When used to represent data in an asynchronous communication scheme, the absence of a neutral state requires other mechanisms for bit synchronization when a separate clock signal is not available.

Figures 8.2, 8.3, 8.4 and 8.5 illustrate NRZ-OOK, RZ-OOK, NRZ-DPSK and RZ DPSK, respectively, from Zhang [1]. Included in these figures are example power spectrum plots [1]. Unity quantum assumed for this discussion.

As noted earlier, OOK is the simplest form of amplitude-shift keying (ASK) modulation that represents digital data as the presence or absence of an optical signal pulse. Of the techniques found in Figures 8.2 and 8.3, NRZ-OOK has the most compact spectrum and the simplest configuration of transceivers [1]. However, it has poor tolerance to dispersion and nonlinearities, in general. On the other hand, RZ-OOK has a shorter signal width than its bit period and an improved tolerance to nonlinearities because of its regular RZ signal pattern [1].

In the DPSK, an optical pulse appears in each bit slot, with the binary data encoded as either a 0 or π optical phase shift between adjacent bits [2]. As one can see from Figures 8.4 and 8.5, the optical power in each bit can occupy the entire bit slot (NRZ-DPSK) or can appear as an optical pulse (RZ-DPSK). The most obvious benefit of DPSK when compared to OOK is the ~ 3 dB lower optical signal-to-noise ratio required to reach a given BER. For example, at a BER of 10^{-9}, the quantum limit for an optically preamplified receiver is 38 photo-electrons per bit for OOK signals [3–5], but only

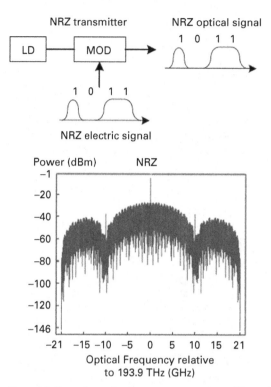

Figure 8.2 Overview of non-return-to-zero on-off key signal modulation scheme used in a fiber optic system [1].

20 photo-electrons per bit for DPSK signals using balanced detection [4–6]. This equates to a ~ 3 dB advantage for DPSK. This clearly can be seen using Figure 8.6, which shows the signal constellations for OOK and DPSK. For the same average optical power, the symbol distance in DPSK (expressed in terms of the optical field) is increased by $\sqrt{2}$. Therefore, only half the average optical power should be needed for DPSK as compared to OOK to achieve the same symbol distance [9]. Xu *et al.* showed this ~ 3 dB advantage in a different manner, plotting $10 \log_{10}(\text{BER})$ versus optical signal-to-noise ratio; see Figure 8.7 (OSNR is equivalent to our SNR_0 used in Chapter 6) [7].

Another scheme employed is carrier-suppressed return-to-zero (CS-RZ), where an optical signal's intensity drops to zero between consecutive bits (RZ), and its phase alternates by π between neighboring bits [1]. Figure 8.8 illustrates CS-RZ and an example power spectrum plot. CS-RZ can be used to generate specific optical modulation formats, e.g., CS-RZ-OOK, in which data are coded on the intensity of the signal using a binary scheme (presence of an optical pulse = 1; absence of an optical pulse = 0), or CS-RZ-DPSK, in which data are coded on the differential phase of the signal, etc. A CS-RZ signal has a spectrum similar to that of an RZ signal, but its frequency peaks (still at a spacing of DR) are shifted by DR/2 with respect to that of RZ, so that no peaks are present at the carrier. The power is ideally zero at the carrier frequency (hence the name). Compared to standard RZ-OOK, the CS-RZ-OOK is considered to be more tolerant to filtering and chromatic dispersion, thanks to its narrower spectrum.

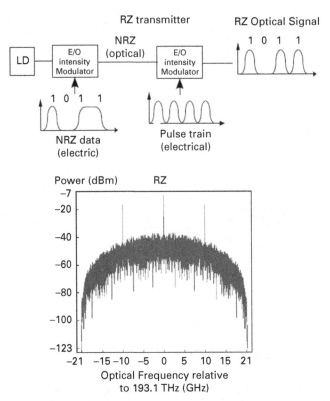

Figure 8.3 Overview of return-to-zero on-off key signal modulation scheme used in a fiber optic system [1].

Many FSOC and Telcom systems are looking at quantum-limited, optically preamplified DSPK operation because it is both energy and spectrally efficient. Caplan cites its Shannon-limited SNR performance to be around 3 ppb, and a spectral efficiency equal to 0.5 bps/Hz ($\eta = 1$) [8, pp. 111–112]. He also pointed out that high-rate, optically preamplified receivers using 24.6% and 7% low-overhead FEC with 0.8 and 0.935 bps/Hz efficiencies have demonstrated 7 and 9 ppb receiver sensitivities at 10 and 40 Gbps data rates, respectively [9–11].

In addition, Caplan points out that many FSOC systems have excess channel bandwidth in spite of spectral efficiency being a primary design parameter [8, pp. 111–112]. The result is improved performance where photo-count efficiency is the design driver. For example, he states the excess bandwidth can be used to improve receiver sensitivity by ~5 dB by using code M-ary orthogonal modulation formats such as M-PPM. Alternative formats can yield improved receiver sensitivity as well. The quantum-limited sensitivity (preamplified, direct detection) for uncoded 1024-PPM, where each symbol carries 10 bits of information, is ~4 ppb (~6 dBppb) ($\eta = 1$), with $10x(M/\log_2 M)$ bandwidth expansion. With the addition of a ~50% overhead (OH) FEC, ~1.5 ppb sensitivity can be achieved, and this can be extended to nearly 1 ppb by implementing optimal soft-decision decoding [12].

Juarez *et al.* discussed the development of a high-sensitivity modem and a high-dynamic-range optical automatic gain controller (OAGC) to provide maximum link

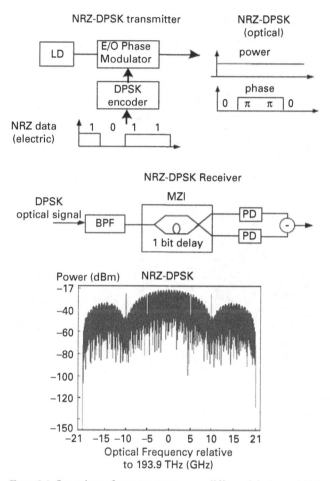

Figure 8.4 Overview of non-return-to-zero differential phase shift key signal modulation scheme used in a fiber optic system [1].

margin and to overcome the dynamic nature of FSOC links [13]. A sensitivity of −48.9 dBmW, i.e., 10 photo-electrons per bit ($\eta = 1$), at 10.7 Gbps was achieved employing an RZ-DPSK-based modem, which includes optical automatic gain control (OAGC) and Reed-Solomon enhanced (255,239) FEC to 7% OH. Low-noise optical gain for the system was provided by an OAGC with a noise figure of 4.1 dB (including system required input loses) and a dynamic range of greater than 60 dB.

8.3   Importance of receiver sensitivity to communications performance

Highly sensitive receivers are key components for the design of high-speed communication FOC and FSOC systems [8, Ch. 4]. They can reduce power-aperture requirements, extend link distances, and/or provide additional link margin to mitigate dynamic channel losses. Their sensitivity depends on coding, modulation and detector quality. It also depends on the optical and electrical filter bandwidths used in the systems. Pfennigbauer

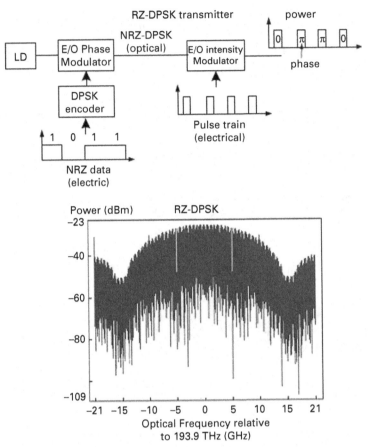

Figure 8.5 Overview of return-to-zero differential phase shift key signal modulation scheme used in a fiber optic system [1].

et al. showed that a direct detection receiver's sensitivity can be optimized by varying both of those filter bandwidths, using the parameter sensitivity parameter γ_p, defined by the equation

$$\gamma_p = 10 \log_{10} \left(n_s / n_q \right) \qquad (8.1)$$

[14]. In the above, n_s is the average number of photo-electrons per bit for a BER = 10^{-9} and n_q is the average number of photo-electrons per bit at the quantum limit, i.e., 38 ppb ($\eta = 1$). Figure 8.9 shows their experimental results for various data rates, R, using the experimental set up in Figure 8.10 and their simulations [14].[1] Here, BERT stands for bit-error-ratio test.

Figure 8.9(a) gives their results for the optical bandwidth [14]. Measurements (bullets), performed for NRZ coding at an electrical bandwidth of $B_e = 0.75R$, lead to

---

[1]   Pfennigbauer *et al.* noted that optical and electrical filter bandwidths involve a careful trade-off between noise on the one hand and, on the other hand, inter-symbol interference (ISI) for NRZ, and peak power reduction due to spectral signal energy truncation for RZ.

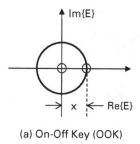

(a) On-Off Key (OOK)

E = Energy Field

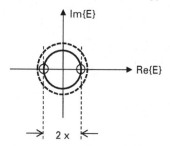

(b) Differential Phase Shift Key (DPSK)

Figure 8.6 Signal constellations for on-off key (OOK) and differential phase shift key (DPSK).

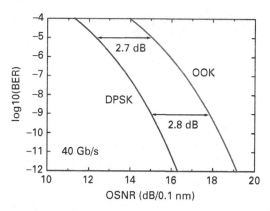

Figure 8.7 BERs versus optical signal-to-noise ratio (OSNR) for DPSK (with a balanced receiver) and OOK in a linear channel, assuming 40 Gb/s transmission [7].

an optimum optical filter bandwidth of $B_0 = 1.35R$. Their corresponding simulations using the advanced Gaussian approximations are represented by solid lines. For RZ coding (measurements: triangles; simulations: dashed lines), they found that an optimum around $B_0 = 2.7R$ is revealed. Here, an electrical bandwidth of $B_e = 0.9R$ was used. The performance gain of RZ compared to NRZ amounts to 1.5 dB, a fact that can mainly be put down to the absence of ISI for the temporally more confined RZ pulses. For both NRZ and RZ, experiment and simulations show excellent agreement. In the

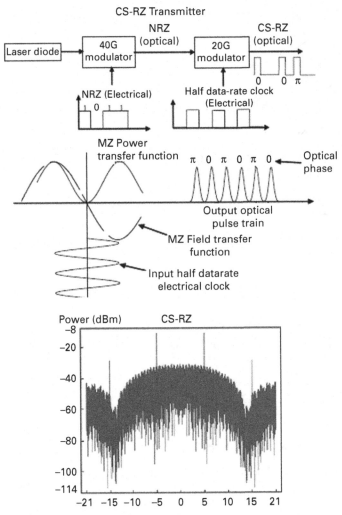

Figure 8.8 Overview of carrier-suppressed return-to-zero (CS-RZ) modulation scheme used in a fiber optic system [1].

case of RZ coding, a sensitivity of 52 ppb ($\eta = 1$) is reached using an optimized optical bandwidth of 3R.

In Figure 8.9(b), Pfennigbauer *et al.* showed the sensitivity as a function of the electrical filter bandwidth for constant optical bandwidth [14]. Again, the bullets and triangles represent experiments, while the solid and dashed lines stand for calculation results for NRZ ($B_0 = 2.86R$) and RZ ($B_0 = 3.12R$) coding, respectively. The thick lines show the calculation results for the advanced Gaussian method, while the thin lines represent the results when applying the standard noise formulas [13]. For NRZ coding, they showed the optimum electrical bandwidth is $B_e = 0.65R$, while for RZ coding the sensitivity is almost independent of the electrical bandwidth when chosen above 0.6R. This can be attributed to the fact that both electrical signal power and σ_{s-sp}^2 are proportional to B_e^2, for $B_e \leq 3R$, which lets their quotient (driving the BER) become independent of B_e.

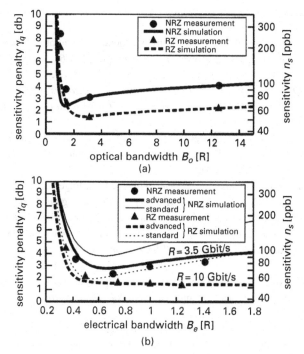

Figure 8.9 Sensitivity penalty relative to the quantum limit as a function of (a) optical filter bandwidth and (b) electrical filter bandwidth. The vertical axis on the right gives the sensitivity in photons per bit (ppb). Measurements (symbols) and simulation (lines) are compared for NRZ coding (solid/bullets) and RZ coding (dashed/triangles) [13].

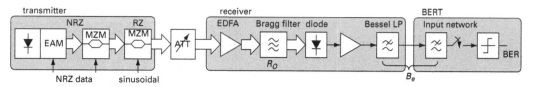

Figure 8.10 Experimental setup: (N)RZ transmitter and optically preamplified direct detection receiver.[13].

References

1. S. Zhang. Advanced optical modulation formats in high-speed lightwave systems. Thesis, The University of Kansas (2004).
2. A. H. Gnauck and P. J. Winzer. Optical phase-shift-keyed transmission. *Journal of Lightwave Technology*, **23**(1) (2005).
3. P. S. Henry. Error-rate performance of optical amplifiers. *In Proceedings of the OFC 1989*, Houston, TX (1989), Paper THK3.
4. P. A. Humblet and M. Azizoglu. On the bit error rate of lightwave systems with optical amplifiers. *Journal of Lightwave Technology*, **9**(11) (1991), pp. 1576–1582.

5. G. Jacobsen. *Noise in Digital Optical Transmission Systems*. Boston, MA: Artech House (1994), Ch. 2.

6. S. R. Chinn, D. M. Boroson and J. C. Livas. Sensitivity of optically preamplified DPSK receivers with Fabry-Perot filters. *Journal of Lightwave Technology*, **14**(3) (1996), pp. 370–376.

7. C. Xu, X. Liu and X. Wei. Differential phase-shift keying for high spectral efficiency optical transmissions. *IEEE Journal of Selected Topics in Quantum Electronics*, **10**(2) (2004).

8. A. K. Majumdar and J. C. Ricklin. *Free-Space Laser Communications; Principles and Advances*. Springer, New York (2008).

9. D. O. Caplan, B. S. Robinson, M. L. Stevens, D. M. Boroson and S. A. Hamilton. High-rate photon-efficient laser communications with near single photon/bit receiver sensitivities. In *Optical Fiber Conference* (OFC), 2006.

10. P. Z. Peeble, Jr. *Digital Communications Systems*. Prentice Hall, Edgewood, NJ (1987).

11. D. M. Boroson. *Optical Communications, A Compendium of Signal Formats, Receiver Architectures, Analysis Mathematics, and Performance Comparison*. MIT (2005).

12. T. Mizuochi *et al.*. Forward error correction based on block turbo code with 3-bit soft decision for 10-Gbps optical communication systems. *IEEE Photonics Technology Letters*, **16** (2004), pp. 1579–1581.

13. J. C. Juarez, D. W. Young, J. E. Sluz and L. B. Stotts. High-sensitivity DPSK receiver for high-bandwidth free-space optical communication links. *Optics Express*, **19**(11) (2011), pp. 10789–10796.

14. M. Pfennigbauer, M. M. Strasser, M. Pauer and P. J. Winzer. Dependence of optically preamplified receiver sensitivity on optical and electrical filter bandwidths – measurement and simulation. *IEEE Photonics Technology Letters*, **14**(6) (2002), pp. 831–833.

9 Light detection and ranging

Lidar (light detection and ranging) is an optical remote sensing technology that measures properties of scattered and reflected light to find range and/or other information about a distant target. The common method to determine distance to an object or surface is to use laser pulses, although as in radar it is possible to use more complex forms of modulation. Also as in radar technology, which uses radio waves instead of light, the range to an object is determined by measuring the two-way time delay between transmission of a pulse and detection of the reflected signal. In the military, the acronym Ladar (laser detection and ranging) is often used. Lidar has also expanded its utility to the detection of constituents of the atmosphere. It is not the intention of this book to expand on all the applications for which this technology has been applied, they are too numerous. Rather we will try to group applications by the technology needed. Thus, for example, a laser speed gun used for traffic monitoring uses the same basic technology as those used for surveying and mapping; i.e., bursts of nanosecond pulses. The wavelengths may vary due to eye safety requirements (>1.5 μm), and the pulse energies may vary because of the distances involved, but both use time of flight measurements. Also the scanning requirements can vary because of the areas to be covered, and the scanning equipment that is available can vary because of the pulse energies involved. For high-energy pulses reflective mirrors are needed, whereas for low-energy pulses electronic scanning with a CCD or CMOS shutter can be used. Most applications use incoherent optics. In some cases, where the additional cost warrants it, coherent (optical heterodyne) detection is used. Such an example is the measurement of clear air turbulence where coherent 10 μm laser ranging systems have been employed. Another class of incoherent applications use backscatter signatures to investigate chemical constituents, primarily in the atmosphere. Such systems use Raman, Brillouin, Rayleigh, and Mie scattering, as well as various types of fluorescence. As we discussed in Chapter 5, Mie theory covers most particulate scattering, ranging from the Rayleigh range $\sim 1/\lambda^4$ for smaller particles (molecules) to the larger particles (aerosols). Raman and Brillouin scattering both induce wavelength shifts, requiring multiple wavelength generation and discrimination. The selection of interference filters, prisms, gratings or spectroscopic machines is determined by the resolution and costs associated with the specific application. It is also possible to combine some of these technologies, such as determining the range, intensity and extent of the chemical constituents. A good reference for available hardware and systems is [1], and for theory [2]. However, a complete discussion of atmospheric research is beyond the scope of this book.

Because the GPS system has become so ubiquitous to ranging systems, and because the two authors of this book developed the first miniaturized version while at DARPA (mini-GPS), we feel it appropriate to digress and explain some of the fundamental properties of the system. The GPS is a spherical ranging system, that is it geo-locates by solving for the intersection of three intersecting spheres. Each satellite (there are 24) has an atomic clock on board which is continuously updated to within a nanosecond of the others. When a receiver on the ground receives the signal from at least four satellites, it first decodes the ephemeris of each satellite (these data are also continuously updated). It then solves for four variables, x, y, z and t. Knowing the reference time and satellite position, it then computes the time of flight and hence the radius and origin of each sphere, from which it solves for the position of the receiver. The signal design is also unique. The basic chip rate is 10.23 MHz so that each chip is 100 ns in duration. There are two overlapping pseudorandom sequences, one at 1.023 MHz and a second at 10.23 MHz. The first sequence (the short code) repeats every millisecond and is for general use. The second sequence (the long code) is encrypted and can only be decoded with a special decoder. The two frequencies selected, L1 = 1575.42 and L2 = 1227.60, are multiples of the chip rate, 154 and 120 respectively. The two frequencies were selected to solve for the frequency-dependent path delays through the atmosphere so that accurate measurements could be made for military use. However, differential-GPS accomplishes the same thing with one frequency by having a surveyed reference station measure pseudo-ranges, which can be sent to the receivers as a vector offset to correct for the atmosphere. The correlators in the receiver can resolve timing to a fraction of the chip duration. In addition, because the chip rates are submultiples of the carrier frequency and bi-phase coded to coincide with the carrier phase transitions, measurement accuracy can be improved to a fraction of the clock cycle with more complex processing that aligns the detected chip transitions with the carrier phase transitions. This allows the carrier to be a fine vernier and is how centimeter and sub-centimeter accuracy is obtained.

9.1   General lidar ranging

We start with the range equation [3]

$$P_{\text{target}} = \frac{P_t \sigma}{\Omega_s R^2} \tag{9.1}$$

to compute the power incident on the area A_t, which we assume to be the area from which the reflection will occur. This is called the target cross section, σ [3]. The return from an isotropic target subtends the solid angle $\sigma/4\pi R^2$ back to the transmitter

$$P_{return} = P_{target} \frac{\sigma}{4\pi R^2} = \frac{P_t}{\Omega_s R^2} \frac{\sigma}{4\pi R^2} \tag{9.2}$$

The transmitter is usually described by its beam width ϕ, hence for narrow beams the solid angle is $\Omega_s = \pi \phi^2/4$. The total return is then

$$P_{return} = \frac{4P_t A_r}{\pi\phi^2 R^2} \frac{\sigma}{4\pi R^2} \tag{9.3}$$

where A_r is the collecting area of the receiver optics. This is the radar range equation. For a purely isotropic return we set $\sigma = A_t$. In optics we are always concerned about what the surface texture of the target is relative to the wavelength. Almost always we can assume that the surface roughness is greater than the wavelength, and hence is a Lambertian reflector. For this case we know (Chapter 2) that the return will be $A_t \cos\theta/_\pi$, which we use to replace $\sigma/_{4\pi}$. This yields

$$P_{return} = \frac{4P_t A_r}{\pi\phi^2 R^2} \frac{A_{target}\cos\theta}{\pi R^2} \tag{9.4}$$

as the return from a Lambertian surface. If the target is smooth, such as a mirror or a corner cube, we will get an enhanced return, or glint, from the target.

Since the lidar measures the two-way propagation time, the distance from the reflector is one half the speed of light multiplied by the measured round trip transit time. Using this timing information it is also possible to eliminate obstacles like foliage, and recreate unobstructed images [1] when multiple angular measurements are available.

Example 9.1 The power on a target P_{target}, having a cross section σ, from a collimated optical source with power P_t and solid transmitting angle $\Omega_s = \pi\left(\frac{\phi R}{2}\right)^2$, can be written as Eq. (9.1). If the target is a mirror of area A_t, and the distance R is great enough so that there is a plane wave incident on the target, the return, Figure 9.1, would reflect into the solid angle, diffraction-limited by the target area, $\Omega_{Reflected} = \lambda^2/_{A_t}$. The power reflected to the transmitter's receiver, with area A_r, becomes

$$P_{Receiver} = \frac{P_t A_r}{\Omega_s R^2} \frac{A_t}{\Omega_{Reflected} R^2} = \frac{P_t A_r}{\Omega_s R^2} \frac{A_t}{\left(\lambda^2/_{A_t}\right)R^2} = \frac{P_t A_r}{\Omega_s R^2} \frac{A_t^2}{\lambda^2 R^2} \tag{9.5}$$

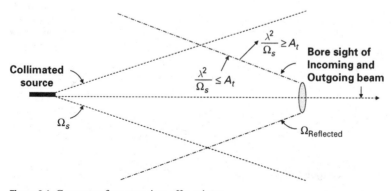

Figure 9.1 Geometry for scattering off a mirror.

Setting this equal to the equivalent isotropic cross section, we have

$$\frac{\sigma}{4\pi R^2} = \frac{A_t^2}{\lambda^2 R^2} \tag{9.6}$$

or the effective cross section of a mirror is

$$\sigma_m = \frac{4\pi A_t^2}{\lambda^2} \tag{9.7}$$

Thus we see that a mirror in space alignment with the boresight of the incoming beam will reflect a cross section shown in Eq. (9.7). This result can be interpreted as the cross section of the actual area multiplied by the gain of an equivalent-sized antenna. In radar this is called a specular reflection, in optics it is called glint. For lidar systems operating at 1 μm, a facet with a 1 mm dimension on a diamond could have a cross section of 4π square meters, albeit only over a 1 mrad beamwidth.

Example 9.2 The effect of a mirror can be simulated by the use of a perfect corner cube [3, p. 27.16]. The geometry is shown in Figure 9.2 and discussed in Appendix C. A ray enters the corner cube parallel to the bore sight, which is perpendicular to the open face of the corner cube. Such a ray will undergo three 45-degree reflections (one off each of the three sides) and will then exit the corner cube going in the opposite direction from whence it came. The effective area is only one-third of the open area of the corner cube due to the limitation given by diffraction theory. In reference 2 this area is given as 0.289 times the square of the side of a cube. Actually it is 0.289 times the square of the open side of the cube. Since the open side of the cube is $\sqrt{2}$ times the side of the cube, the area would actually be 0.579 times the square of the side of the cube.

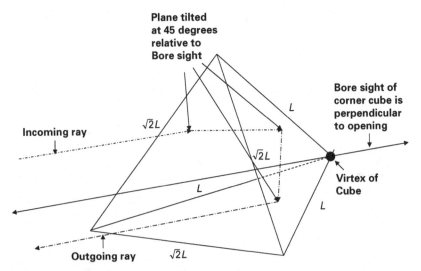

Figure 9.2 Geometry for scattering off a corner cube.

In addition, each of the three sides reflects independently with the potential for power combining.

What we have shown is that the reflection from a surface is greatly influenced by the correlation lengths of the imperfections as well as the vertical depth. In optics, because of the shortness of the wavelength, glint can be created by small reflecting targets with sheared edges, such as sand. An interesting evaluation of this topic was given by Ruze [4], when discussing the limitations in antenna gain in a reflecting structure due to surface imperfections. He showed that a surface with dimension D having imperfections with a correlation length $C_{corr} < D$ had a gain equivalent to a surface with dimension C_{corr}. This gain was further reduced if these imperfections had an additional superimposed randomness (Figure 9.3). In an optical surface these are referred to as "flatness" (C_{corr}) and "roughness". High-quality optical surfaces are generally flat and smooth to $\lambda/20$ over the entire surface. We saw earlier, Eq. (2.39), that if the roughness had an impulse-like correlation, the receiver recreated the object image in the focal plane. This occurs because broadband light does not create regions of constructive and destructive interference. What we observe is then the actual

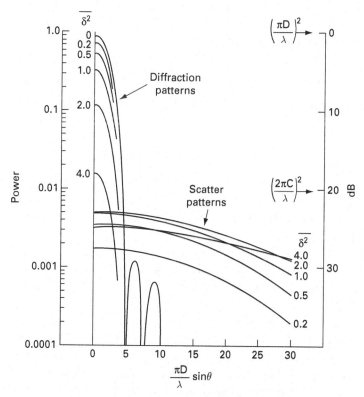

Figure 9.3 Radiation patterns of phase distorted circular aperture, 12 dB illumination taper, $D = 20c$ [4].

shadowing of the object. If the incident light is narrowband as in a laser, the texture becomes evident and appears as "speckle". This is due to the creation of a random pattern of constructive and destructive interference [5].

Example 9.3 Consider a one square inch sheet of Scotchlite with 17MIL crystals (Appendix C; Eq. (C.2)). The reflection from a headlight (0.5 μm incoherent) would have the equivalent reflection of a 1.146m^2 Lambertian surface (Figure 9.4), assuming a unity fill factor.

We next show how signal-to-noise ratio calculations are made; Example 9.4.

Example 9.4 Compute the signal-to-noise ratio from the return off a corner cube.

The cross section, σ_{cc}, of a corner cube is

$$\sigma_{cc} = \frac{4\pi A_{cc}^2}{\lambda^2} : \quad A_{cc} = 0.579L^2 \tag{9.8}$$

which is the effective area radiating back to the transmitter and L is the side of the cube. The power reradiated from the target is

$$P_{target} = \frac{P_t \sigma_{target}}{\Omega_t R^2} = \frac{4P_t}{\phi^2 \pi R^2} \frac{4\pi A_{cc}^2}{4\pi R^2 \lambda^2} \tag{9.9}$$

The received power becomes

$$P_{return,_{cc}} = \frac{4P_t}{\phi^2 \pi R^2} \frac{A_{cc}^2}{R^2 \lambda^2} A_t \tag{9.10}$$

The signal-to-noise ratio of a lidar system, with pulse width τ ($\tau = 1/2B$), can be calculated as

$$SNR = \frac{\eta P_{return}^2}{(P_{return} + P_{noise})2hfB} = \frac{\eta k_{return}^2}{k_{return} + \frac{P_{noise}\tau}{hf}} \tag{9.11}$$

where we have assumed a pulsed system with photo-electron counting, and observe the signal-to-noise ratio in the interval τ. In most cases the noise count is less than one and can be ignored. In Figure 9.4 we show the ratio of the return off a corner cube to that off an isotropic target with the same effective area, in dB. Notice the dramatic difference, which is why glint off even a miniscule target can be large. However, there is an area of partial coherence between a perfect mirror (corner reflector) and an isotropic surface.

Notice that when the roughness is 0.3 μm, or roughly λ/5, the corner cube appears isotropic. In optics there is also the component of "flatness". This would correspond to the case where the "roughness" had a large coherence length. In general, both components are needed to specify the optical surface. If the uncorrelated portion of the optical surface, the roughness, is greater than 10 μm, it doesn't matter what the

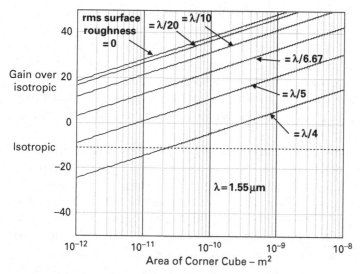

Figure 9.4 Gain of a corner cube over isotropic.

correlation length is. In antenna design, the primary interest is in flatness [4], defined by a correlation length C (see Figure 9.3).

On a final note we should point out that even though we have considered a normal atmospheric channel, there will be backscatter from particulates that make it up. This return will consist of an exponentially decaying background with the target return riding on it. Thus as in other radars some sensitivity time control (overlap function [2]) will be necessary. For the lidar, target extraction usually consists of a leading edge detector or, for a repetitive signal, a leading edge tracker. This is akin to taking the derivative of the return, which accentuates the sharp rise in the return pulse. If the receiver has accurate knowledge of its own position, it can be transferred to the target by having accurate angle and range information from the target return. In addition, the receiver must have accurate knowledge of its orientation. If the receiver is ground-based this can be obtained by an inclinometer. If the receiver is on a moving platform this must be obtained from the vehicle's inertial navigation sytem (INS). The general areas of use for this class of lidar include the military, surveying and mapping, law enforcement and hydrology. Extensive geological world maps exist that have been produced with lidar systems.

9.2   Single scattering (backscatter) applications

What we showed in section 9.1 applies to all lidar systems that mimic their radar counterparts. And while the specific components may look somewhat different, the functions are identical. In this section, which is really the diagnostic set of applications, we see the broader range of uniquely lidar capabilities. Hence what was the interference in section 9.1 is now the signal. If, for example, this has a Raman-effect-generated return, we would be interested in the character of the return and also where it occurs in time, which translates to where it is in space. We might also be interested in relative intensities as well as relative frequencies. To be useful we need to quantify a calculation for the return, hence we need to know the backscatter from the atmosphere. A detailed discussion of this is given in [6], where Rayleigh scattering (elastic scattering) is assumed and, using the properties of a perfect gas, the results are applied to transmission through a vertical atmosphere.

To be absolutely accurate in modeling the atmosphere is difficult because of the variability of the pressures and temperatures. Furthermore equipment can't always be built better, only calibrated better. This in fact is why good experiments are made to yield differential results and avoid the need for absolute calibration [2, p, 331; 6]. For this reason we will follow the lead given in [7] and use basic physical quantities for a first-level design. We justify this by reference to [8] where 30 years later it was pointed out that more accurate data would have improved the measurements by 5–10%.

For the diagnostic lidar, the identification of the constituents along the propagation path is the primary concern. Therefore the radar range equation takes the (range-corrected) form (from Eq. 9.3)

$$z^2 E_{return} = \frac{4E_t A_r}{\pi \varphi^2} \frac{\sigma_b}{4\pi} e^{-2\int_0^R c(z)\,dz} = K\sigma_b e^{-2\int_0^R c(z)\,dz} \tag{9.12}$$

The term $\sigma_b = \sigma_s (\theta = \pi)$ is the backscatter volume cross section and $c(r)$ is the volume extinction coefficient. By ensuring a highly collimated transmit beam, and a receiver field of view that always subtends the common scattering volume, the geometric loss on transmission can be essentially eliminated. A range-normalized version of this range equation is also used which is

$$L(z) = \ln(z^2 E_r(z)) \tag{9.13}$$

If we let z_0 be the reference range then

$$L(z) - L(z_0) = \ln\left(\frac{\sigma_b}{\sigma_{b0}}\right) - 2\int_{z_0}^z c(z)\,dz \tag{9.14}$$

It can also be put into differential form

$$\frac{dL}{dz} = \frac{1}{\sigma_b(z)} \frac{d\sigma_b(z)}{dz} - 2c(z) \tag{9.15}$$

Example 9.5 Assume that the scatterers are homogeneously distributed in space such that

$$\frac{d\sigma_b(z)}{dz} = 0 \tag{9.16}$$

This would then yield

$$\frac{dL}{dz} = -2c(z) \tag{9.17}$$

and only holds for a homogeneous path.

An expression for the Rayleigh volume backscatter from a gas was developed in [7]. In general there will be a relationship between the pressure, volume and temperature of molecules in the atmosphere

$$P(z)V = nRT(z) \tag{9.18}$$

and the scaling law behaves as

$$\frac{P(z)}{P_0} \cdot \frac{T_0}{T(z)} \Rightarrow P(z) = \frac{P_0 T_0}{T(z)} \tag{9.19}$$

Example 9.6 We start with the perfect gas law that relates pressure P and temperature T in the atmosphere by

$$P(z) = P_0 e^{-\frac{gz}{RT}} \tag{9.20}$$

where P_0 is the pressure at zero altitude, $R = 2.8706 \times 10^2 \, \mathrm{J\,kg^{-1}\,K^{-1}}$ is the universal gas constant and g is the gravitational force ($6.67384 \times 10^{-11} \, \mathrm{m^3 kg^{-1}\,s^{-2}}$). The temperature is generally written as $T = T_0 - \gamma z$ where T_0 is the temperature at zero altitude, and set the lapse rate γ equal to zero. This is the so-called isothermal assumption. The total scattering cross section of an isotropic molecule illuminated by polarized light can be written as [7]

$$\sigma_b = \frac{8\pi^3 \left(n^2 - 1\right)^2}{3N\lambda^4} \tag{9.21}$$

where n is the index of refraction of air, N is the number density of air molecules and λ is the wavelength. The volume scatter coefficient then becomes

$$\sigma_b(z) = \frac{8\pi^3 \left(n^2 - 1\right)^2 P(z)}{3N\lambda^4} = \frac{8\pi^3 \left(n^2 - 1\right)^2 P_0 e^{-\frac{z}{H}}}{3N\lambda^4} \tag{9.22}$$

where $H = RT_0 / g$ is the scale height. From the US Standard Atmosphere [9], we take at zero altitude standard constants and scale pressure

$$\begin{aligned}
T_0 &= 288.80^\circ \mathrm{K} \\
P_0 &= 1.013 \times 10^3 \, \mathrm{mb} \\
N &= 2.55 \times 10^{25} \, \mathrm{m^3} \\
n &= 1.000293
\end{aligned} \tag{9.23}$$

This is called Rayleigh or elastic scattering. From Eq. 9.15 we obtain

$$\frac{dL}{dz} = \frac{1}{\sigma_b(z)} \frac{d\sigma_b(z)}{dz} - 2c(z)$$

$$= -\frac{1}{H} - 2\sigma_b(z) \tag{9.24}$$

The actual extinction coefficient is related to the logarithm of this return, hence must be corrected for any secondary generated forward scattering by the return. Also correction terms for polarization and lapse rate can be included, but are part of the scientific interpretation and are not covered here. There are other atmospheric models, constant density and polytropic, which are not discussed here since they are only instructive but violate some physical laws.

An empirical model for the lidar ratio was introduced by Curcio and Knestric [10] while investigating the relationship between atmospheric extinction and backscatter. While later investigators have written about the limitations of this model [2], it nevertheless has been used extensively in the literature [11,12]. It takes the form

$$\sigma_a(\theta = \pi) = B_1 \sigma_m{}^{b_1} \tag{9.25}$$

and can be applied in the single-component case. For this case we have the equation

$$[Z(r)]^{1/b_1} = [KB_1]^{1/b_1} c(r) e^{-\frac{2}{b_1} \int_{r_0}^{r} c(z) dz} \tag{9.26}$$

which can be re-written as [11]

$$c(r) = \frac{[Z(r)]^{1/b_1}}{[KB_1]^{1/b_1} - \frac{2}{b_1} \int_{r_0}^{r} [Z(x)]^{1/b_1} dx} \tag{9.27}$$

Multiple components

We have only been considering the single-component solution, namely molecules. An important metric is the lidar extinction to backscatter ratio. This ratio is important for the more detailed calibration of the atmosphere. It is obtained by recognizing that both aerosols and molecules exist in the atmosphere and introduces the two-component solution. The total range equation becomes

$$E_{return} = E_0 \frac{A_r}{2r^2} \left[\frac{3}{8\pi} \sigma_{bm}(z) + \frac{\sigma_{ba}(z)}{4\pi} \right] e^{-2\int_{0}^{R} c(z) dz} \tag{9.28}$$

where we have assumed that multiple scattering and any background signal are negligible. We have also included the aerosol backscatter coefficient from Mie theory and the total extinction coefficient $c(z)$. The range is determined by $z = ct/2$, where t is the time between transmission and reception of the laser pulse. This is separated into two equations, one for molecules

$$E_m = E_0 \frac{A_r}{2z^2}\left[\frac{3}{8\pi}\sigma_{bm}(z)\right]e^{-2\tau(z)} \tag{9.29}$$

and one for the aerosols,

$$E_a = E_0 \frac{A_r}{2r^2}\left[\frac{\sigma_{ba}(z)}{4\pi}\right]e^{-2\tau(z)} \tag{9.30}$$

with

$$\tau(z) = \int_0^z c(z)dz \tag{9.31}$$

the optical thickness. The two constituents are separated by observing the spectrum generated by Raman (inelastic) scattering. The scattered returns are also Doppler-shifted due to the wind motion. However, the heavier aerosols have a much smaller Doppler spread than the lighter molecules (O_2 and N_2); hence they can be separated in the spectrum (Figure 9.5) using a narrowband filter. Having a second channel to monitor the scattered return at the radiated frequency allows for calibration.

The molecular scattering is proportional to the pressure, Eq. (9.18). Then from Eq. (9.14) we have as a ruler

$$\tau(z)-\tau(z_0) = \frac{1}{2}\ln\left(\frac{P(z)}{P(z_0)}\right) - \frac{1}{2}\ln\left(\frac{L_m(z)}{L_m(z_0)}\right) \tag{9.32}$$

An average ratio (lidar ratio) can also be obtained as

$$R(z) = \frac{E_a(z)}{E_m(z)} = \frac{2\sigma_a(\theta = \pi)}{3\sigma_m} \tag{9.33}$$

from which the aerosol backscatter is obtained from the molecular backscatter as

$$\frac{\sigma_a(\theta = \pi)}{4\pi} = R(z)\sigma_m(z)\frac{3}{8\pi} \tag{9.34}$$

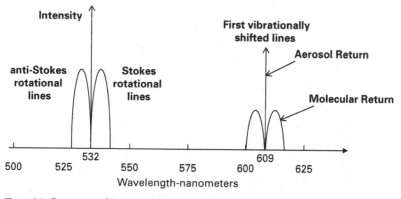

Figure 9.5 Spectrum of Raman shifted return from lidar.

$(\sigma_m(z)$ is isotropic). Other corrections are also made for polarization and non-spherical particles. Corrections for temperature can be made using the Atmospheric Emitted Radiance Interferometer (AERI) [13] and radiosonde. Vertical profiles can be generated using $\tau(z) - \tau(z_0)$.

Multiple wavelengths are usually employed to aid in the analysis. Since there is no true analytic description of a particle size distribution, using the fact that Mie theory is wavelength-dependent offers a means of obtaining some characteristics of the actual aerosols present. For example we know in the Rayleigh region the scattering efficiency behaves as $1/\lambda^4$, whereas in the large-particle region the scattering efficiency approaches twice the geometric cross section of the particles. One option is to try to invert b in Eq. (5.8). This is not only difficult, but is further complicated by the lack of knowledge of the particle index. The alternative most commonly used is to try a variety of lidar wavelengths and some a priori assumptions about the media to extract usable metrics.

Range height indicator

An approach to multiple angle measurements is called RHI scanning. In this technique the lidar scans the atmosphere at a constant azimuth, starting at near horizontal and scanning to the zenith (Figure 9.6). This technique is described in detail in [2]. It assumes a horizontally layered atmosphere, with each of the layers having a homogeneous distribution of scatterers during the period of probing. Although one can in theory require only two angles, the fact that the data are inherently noisy warrants a multi-angular probing. This technique is most applicable during stable atmospheric conditions, which usually occur at night.

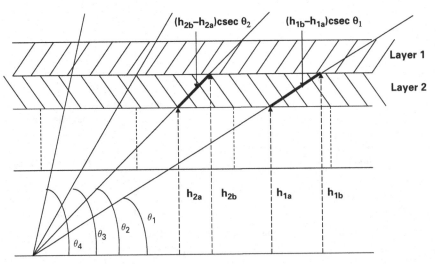

Figure 9.6 Lidar multi-angular measurements.

It is not possible to truly cover the broad breadth of lidar applications in this book. The general areas of use for diagnostic lidar include agriculture, oceanography, archeology, conservation, hydrology, physics, astronomy, and probably many more. Examples of Raman-shifted spectra can be seen in [14]. A sampling of some of these applications is shown only to emphasize the range of usage [15–21]. There is also an area of bistatic lidar which is not covered here [22–24], but is commented on in section 9.5.

9.3 Ranging in a multiple scattering environment

There is no single model for pulse transmission in a multiple scattering environment. We do know that there exists an unscattered component which propagates as though the media were absent, but decays as e^{-cz}. In Chapter 5 we used this result to demonstrate how we could extract propagating images embedded in a scattering environment such as water. We also have a good model for the propagation of the scattered beam past one scattering length when there is also absorption. This is the forward angle scattering model, which works because rays propagating off axis experience greater losses than the on-axis rays, hence contribute little. This is what causes the "manhole" effect when viewing the Sun from below the water. At sufficient depths rays coming from the vertical are minimally absorbed, hence that is where the light seems to appear. Monte Carlo routines also model the multiple scattering environment, but cannot propagate an image. It is possible to use the forward scattering approximation (Appendix B) to simulate the response to a finite pulse in the scattering environment. This spatial impulse response point spread function can be written as

$$E(z,r) = \frac{e^{-a\left[z^2+r^2\right]^{1/2} - \frac{r^2}{2/3\, b\Omega_{rms}\left[z^2+r^2\right]^{3/2}}}}{2/3\, b\Omega_{rms}\left[z^2+r^2\right]^{3/2}} \tag{9.35}$$

where it is assumed that all angles are collected. This overstates the rays collected, which can be reduced with the inclusion of a spatial filter. As with the lidar equation this is a spatial response that is converted to time using the relationship $z = ct/2$. In the lidar case the receiver had a narrow field of view and attempted to only view the single scattered return. This cannot be done in the multiple scattered case because of the dominance of the multiple scatter. Instead we create a virtual scattering layer as in Figure 9.7. We then assume that for the multiple scattering case, the reflection coefficient ρ_0 represents the fraction backscattered from that layer as a mirror image of what continues in the forward direction (First Law). For the monostatic geometry we consider the scatter from a ring at the depth z, of thickness Δz with volume $2\pi\Delta r\,\Delta z$, to arrive at the point z at the time

$$t = \frac{\sqrt{z^2+r^2}}{c/1.33} \tag{9.36}$$

and to the origin at the same time. The scatter from the virtual layer is estimated as

$$E(z)-E(z+\Delta z) \tag{9.37}$$

The backscattered contribution from each element of the layer is attenuated by the backscatter coefficient ρ_0. We select Δz for the time resolution required. Since the index

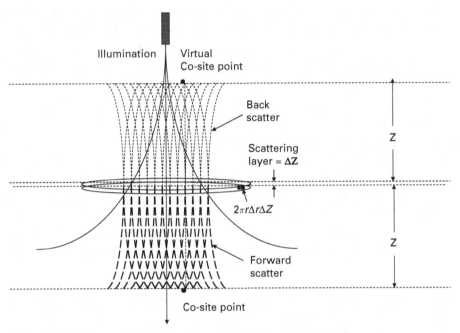

Figure 9.7 Model for modelling pulse response in multiple scattering media.

in water is 1.33, light propagates at 0.225 meters per nanosecond. Setting $\Delta z = 0.225$ gives 1 ns increments. In Figure 9.8 we show the results of the scattering as a function of depth, starting at one scattering length equal to 27 meters. We show the total scatter in the forward and backward directions, and the forward scatter from several of the rings. In Figure 9.9 we show an estimate of the pulse shapes at various depths on a logarithmic scale and in Figure 9.10 we plot this on a linear scale to better see the actual pulses. From Figure 9.9 we see that at any depth z, the direct return comes in first, followed immediately by the close in smaller rings. As the rings get larger the signal keeps increasing until the added loss due to distance overtakes the ring growth, and the return starts to decay. We have not included the backscatter, which is only 5% and would not significantly change anything.

From Figure 9.10 we have estimated the half-power points of a few of these shapes and compared them with Stotts's result [25] for the standard deviation,

$$\Delta t = \frac{z\left[\left[\frac{.3\left(\left(1+2.25(sz\Omega)^{3/2}\right)-1\right)}{(sz\Omega)}\right]-1\right]}{c/n} \tag{9.38}$$

Although the general trend is the same (Figure 9.11), the model shows a slower growth rate in the pulse spreading as a function of depth. This could be because the Stotts model does not have absorption in the calculation. As we have discussed earlier, the absorption of the media tends to attenuate the longer rays associated with multiple

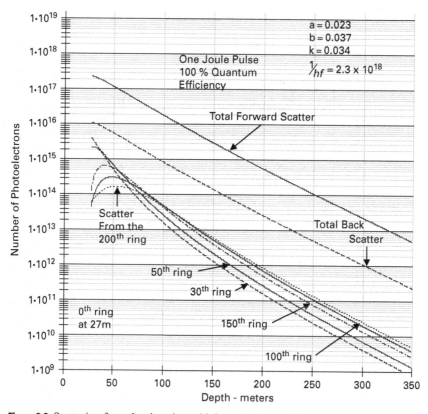

Figure 9.8 Scattering from the rings in multiple scattering model.

scattering. This is why the Stotts model works well in clouds where there is virtually no absorption.

9.4 Ranging in water

In previous sections we presented a model for the illumination in the ocean using a finite beam at the ocean surface (Appendix B). This model was shown to have excellent agreement with the embedded model for distances greater than one scattering length. The model gives the intensity profile of the beam as it propagates to various depths. In the previous section we extended the model to estimate the effects on pulse propagation. In Figure 9.12 we show the geometry of a finite beam illuminating the ocean surface. We draw a hypothetical diffusion layer with both upwelling and downwelling. In fact, this layer exists continuously throughout the water column, and is caused by the scattering, both forward and backward, of the elemental particles that constitute the water column. In pure water these particles are water molecules, which are small and scatter broadly. As contaminants accumulate in the water, they tend to be larger than

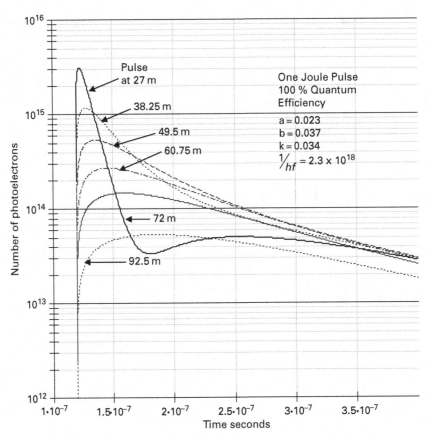

Figure 9.9 Computed pulse responses by depth from multiple scattering model.

water molecules, hence scatter more in the forward direction. In addition to scattering, these contaminants also absorb, creating additional loss in the water. Through this process the water starts out looking pure blue in deep ocean, turning to green and then brown in the littoral regions, where estuaries cause the contamination. In Figure 9.13 we show measured data of this reflectance as a function of wavelength, for various water types (reflectance is defined as the ratio of upwelling to downwelling). The numbers shown represent the values at 515 nm, where all water types are the same, 1.8%. For the water types we encounter, deep ocean (Ib) and littoral (II), the values are higher, 4.3% and 3.2% respectively. The upwelling at 455 nm is 7%. When we compute the noise reflected from the ocean by a continuous source such as the Sun, we use the upwelling number.

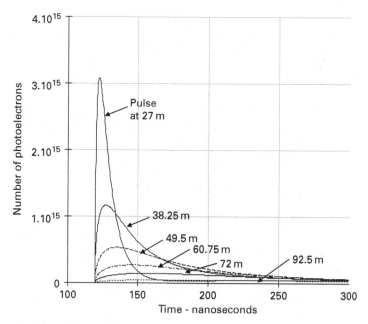

Figure 9.10 Computed pulse responses by depth from multiple scattering model on linear scale.

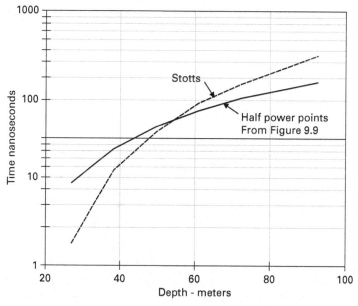

Figure 9.11 Comparison of half power point estimate of model to variance in Stotts model of pulse width.

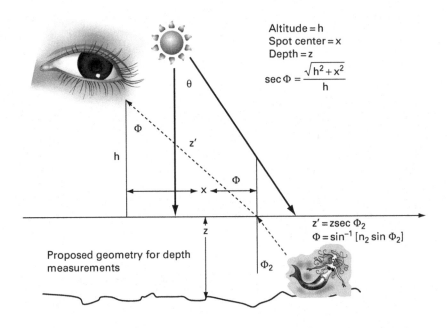

Altitude = h
Spot center = x
Depth = z

$$\sec \Phi = \frac{\sqrt{h^2 + x^2}}{h}$$

$z' = z\sec \Phi_2$
$\Phi = \sin^{-1}[n_2 \sin \Phi_2]$

Proposed geometry for depth measurements

Figure 9.12 Geometry of imaging an underwater object.

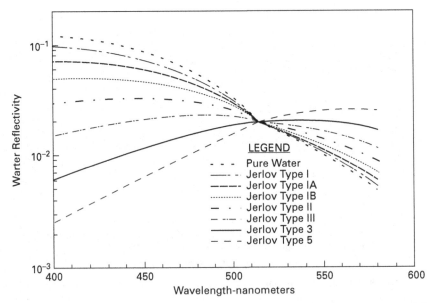

LEGEND
- - -   Pure Water
— · —   Jerlov Type I
— — —   Jerlov Type IA
············   Jerlov Type IB
— · —   Jerlov Type II
— ·· —   Jerlov Type III
———   Jerlov Type 3
— — —   Jerlov Type 5

Figure 9.13 Water reflectivity ρ as a function of wavelength for Jerlov water types I, II, III, 3, and 5.

The illumination

Because of the continuous backscatter, the return of a pulsed laser will show an exponential decrease in time as the return reflects back from the continuous water column. At every layer, including the one where the object is located, the scattering profile from the elemental particles is a figure eight, with the forward circle larger than the return circle, reflecting the difference in scattering level. Still these scattering profiles are broad angle, both in the forward and backward directions. As discussed in the previous section we can model the diffuse return from the depth z, by observing the mirror image of the intensity at the depth $2z$, attenuated by this reflectance. In particular, this would be the return at the surface from the layer where the objects lie. In general, the area covered by the illumination at depth z will be much greater than the area of the object, so to first order we can model this diffuse return by the value without the object. This approximation will be an upper bound, but very tight. A better approximation would be to propagate the light that would have gone through the region where the object was to $2z$, and subtract it from the total. In Appendix B we show how the intensity varies as a function of spot size and depth.

The return

We showed in Chapter 5 that the response to a spot illumination slowly reaches the embedded result as a function of depth. Therefore we would expect the return from a finite target to behave similarly. Without identifying a specific target, it is difficult to make a calculation. However, we showed how to calculate the intensity as a function of depth for different spot sizes. We also showed how to calculate the diffuse return from the water. Once we have a target, we can integrate the reflection from the target and propagate it back to the surface with the same set of equations. We have also shown how the pulse distorts as it propagates.

The received energy

After we have decided on the spot size viewed by the receiver, we must compute the solid angle subtended by the platform to the spot, to determine the background noise incident on the receiver. We must also account for the range loss of the signal to the platform. For this the model predicts that the radiation exiting the ocean is isotropic with a $\cos \theta$ falloff, and a $1/R^2$ range loss, where R is the actual range to the platform. The only assumption not discussed was the nature of the reflectance off the target. We assume that the radiation is isotropic. For a fixed pulse energy, and the same spot size for transmit and receive, there is an optimum spot size. If we make the spot too small, then as the return spreads due to scattering, we only intercept a portion of the return. If we make the spot too large, then the integrated return is too small. To perform the actual calculation would be a six-dimensional integration, which we have not performed. If possible one would optimize the spot size with the target depth to maximize power on target.

9.5   Pulsed imaging

In this section we will discuss the concept of pulsed imaging in a multiple scattering channel, and incorporate some of the ideas from Chapter 5. One of the authors (SK) first became aware of the work of Professor Andrzej Sluzek at Nanyang Technological University, when visiting him in Singapore in 2010. Professor Sluzek had constructed a tank with a simulated multiple scattering environment and made pulsed imagery of submerged targets [26, 27]. He transmitted a very narrow laser pulse (a few nanoseconds) and used a gated camera as a receiver. He gated the returns from the image to coincide with the expected echo. Professor Sluzek was asked whether we could have the raw data from his reconstructed return imagery, to see whether further enhancement to his imagery could be achieved with some of the techniques presented in Chapter 5. He sent the imagery and the ad hoc algorithms were applied to them. The results are shown in Figures 9.14–9.17. Although the data don't show great improvement to his originals, there are reasons for this. First, he is not sub-visible and hence has plenty of signal-to-noise ratio since he can gate out a great deal of the scatter. Second, he used a camera geared for analog TV display, hence he went from an analog focal plane to a digital data

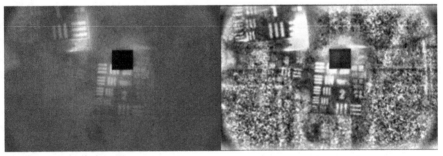

Enhanced Pulsed Imagery

Figure 9.14 Processing of Professor Sluzek's first supplied image (raw image on left, processed image on right).

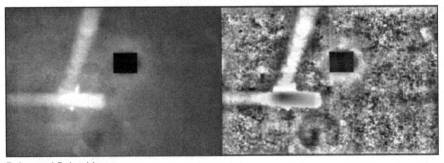

Enhanced Pulsed Imagery

Figure 9.15

Enhanced Pulsed Imagery

Figure 9.16

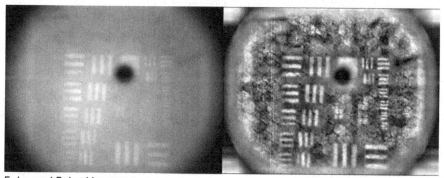

Enhanced Pulsed Imagery

Figure 9.17 Processing of Professor Sluzek's third supplied image (raw image on left, processed image on right).

extraction and then back to the analog output, which was then digitized. As a result the data were corrupted but still showed a potential for adding 2–3 extinction lengths to his capability. The significance of these results lies not only in underwater imaging, but also for pulsed imaging in dust (mines), fog and other deleterious channels at wavelengths more suited to their properties.

Assume that, as in section 9.2, all the power can be focused on the image, only taking the e^{-kz} loss. Since this is now an illuminator, all the power is useful. If we assume we are illuminating only the N pixels, each one will reflect the fraction $1/N$ of the power (Lambertian reflector) and return with an additional loss of $e^{-cz}/N\pi z^2$ The received signal energy per pixel will then be

$$E_s = \frac{E_t e^{-(k+c)z}}{\pi N z^2} \tag{9.39}$$

If one is interested in the signal return from a distance of z meters, with the time equal to $z/$(speed of light $= c$), there will also be latent scatter returns from depths less than z arriving at the same time.

Example 9.7 Compute the signal to noise ratio of a pulsed laser imaging system in Jerlov type Ia water, $a=0.023$, $b=0.037$, $k=0.034$. We assume that a one joule pulse propagates as in Appendix B (equation B.7 and applying reciprocity) and illuminates a target of 1 cm^2 with intensity I. Assume that the return image decays as e^{-cz}/z^2. We also assume that the noise received per pixel N also propagates as in Appendix B, but over the distance 2z. Therefore the signal-to-noise ratio per resolution cell falls off as

$$SNR = \frac{\left[\dfrac{E_t e^{-(k+c)z}}{\pi N z^2}\right]^2}{\rho_o E_t e^{-2kz}\Big/ \pi N} = \frac{E_t e^{-2cz}}{\rho_o \pi N z^4} \tag{9.40}$$

where ρ_o is the diffuse reflection coefficient which is taken to be 0.07. Equation 9.40 is plotted in Figure 9.18. I and N can also be approximated as, $\exp(-kz)$ and $\exp(-2kz)$ respectively. This result is for water and should apply to other media.

Although only qualitative, the model also implies a substantial improvement with bi-static operation. The improvement appears best at shorter separations and shorter ranges, although significant improvement appears under most conditions. It would seem that the multiple scattered returns travel longer distances, hence suffer greater losses, and this holds for Jerlov Type 1a water past 16 meters, and Type II water past 10 meters bi-static offset. Remember the receiver is only gated when the pulse is expected. These qualitative results

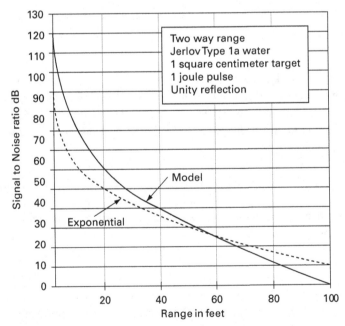

Figure 9.18 Model prediction of signal to noise ratio for pulsed imagery in Type 1a water.

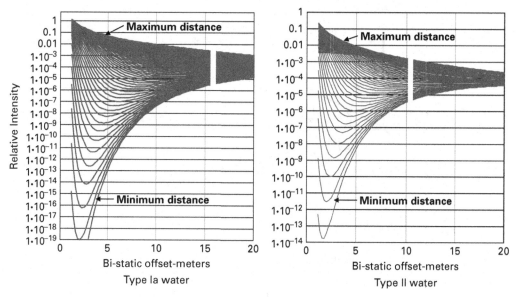

Figure 9.19 Speculation of bi-static improvement for pulsed imagery.

have been plotted in Figure 9.19, with the vertical white box indicating where the limit of the model holds. Nevertheless the fact that the intensity is curving downward at this point leads one to believe that there could be an optimum bi-static offset. This conclusion is also supported by the work in over-the-horizon single scatter communications [27], where the single scatter return drops off rapidly with increased bi-static angle, reaching an asymptotic result of about −30 dB. There is a simultaneous, but delayed, build-up in multiple scattering which peaks and then decays as shown in Figure 9.10. This observation might also be useful to the atmospheric probing in section 9.2.

9.6   Geo-location

We initiated this chapter with a brief description of the GPS system because it has become so intertwined with all forms of ranging systems. It is therefore appropriate to end this chapter with a brief description of geo-location. In its simplest form it consists of projecting a set of geodesic coordinates (in our case a straight line) between a set of measuring equipment and some projected target (Figure 9.20). Starting from the sensor location (x_0, y_0, z_0), we wish to project to the point (x_1, y_1, z_1). The projection is $\vec{R}$.

$$\vec{R} = (R\cos\theta\,\cos\phi,\ R\sin\theta,\ R\cos\theta\,\sin\phi)$$
$$d\vec{R} = \Delta R(-(\sin\theta\,\cos\phi\,\Delta\theta + \cos\theta\,\sin\phi\,\Delta\theta),$$
$$\cos\theta\,\Delta\theta(\cos\theta\,\sin\phi\,\Delta\phi - \sin\theta\,\sin\phi\,\Delta\theta))$$
$$R = |\vec{R}| \tag{9.41}$$

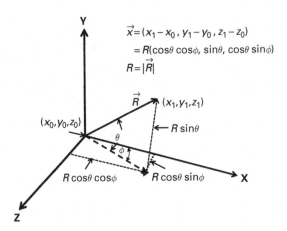

Figure 9.20 Projection of the geodesic.

The error in this projection is $d\vec{R}$ where

$$|d\vec{R}| = \sqrt{\left(\frac{\partial \vec{R}}{\partial \theta}\right)^2 \sigma_\theta^2 + \left(\frac{\partial \vec{R}}{\partial \phi}\right)^2 \sigma_\phi^2 + \sigma_R^2} \tag{9.42}$$

and $\partial\theta$, $\partial\phi$, ∂R are random variables with variances σ_θ^2, σ_ϕ^2, σ_R^2 respectively. We have

$$\left(\frac{\partial \vec{R}}{\partial \theta}\right)^2 = R^2$$

$$\left(\frac{\partial \vec{R}}{\partial \phi}\right)^2 = R^2 \cos^2\theta \tag{9.43}$$

from which we conclude that

$$|error| = \sqrt{\sigma_R^2 + R^2 \sigma_\theta^2 + R^2 \sigma_\phi^2 \cos^2\theta} \tag{9.44}$$

This implies that having GPS and a good laser ranging system is only one part of the equation. The other part of the equation says that as we try to project further out, the accuracy of the gimbals becomes important. Thus in order to maintain a ranging accuracy of 1 foot (0.3 m) at a distance R, requires that

$$\frac{\sigma_R}{R} >> \sqrt{\sigma_\theta^2 + \sigma_\phi^2 \cos^2\theta} \tag{9.45}$$

For R equal to 1 kilometer this is < 300 μrad, whereas for R equal to 10 kilometers this becomes < 30 μrad. Hence even a ground measurement imposes fairly strict requirements on the optical mount. An accurate θ measurement requires an accurate inclinometer, whereas an accurate measurement of φ requires a good compass. A magnetic compass might not suffice at the higher latitudes where it starts to deviate from true north. In addition, in mountainous terrain we are subject to gravitational anomalies which can

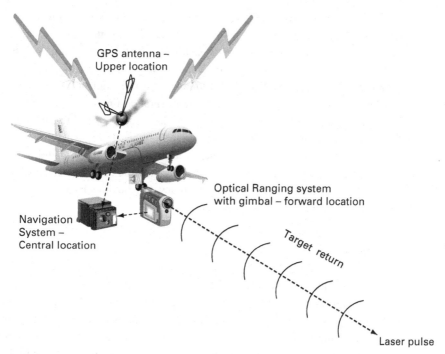

GPS antenna –
Upper location

Optical Ranging system
with gimbal – forward location

Navigation
System –
Central location

Target return

Laser pulse

Figure 9.21 Static offset.

cause serious errors in an inclinometer output. Thus even for the simple case of a ground measurement, good equipment is required.

Now let us consider the case where the sensor equipment is on an aircraft traveling at 200 meters per second. In addition to the requirement for accurate elevation and azimuth measuring equipment previously discussed, our reference coordinates are continuously changing. To complicate the matter further, the GPS system measures the location to the phase center of the antenna which is not co-located with the center of rotation of the sensor gimbal. Finally, the offset lever arm (Figure 9.21), while rigid, is constantly changing orientation. What is needed is an accurate instrument navigation system (INS) which contains an inertial measurement unit (IMU). Furthermore for the greater accuracies, this INS must be navigation-grade. A navigation-grade INS system would have the following specifications:

gyro drift rate	0.01 degrees/hour
accelerometer bias	50–100 mg
attitude accuracy	several µrad
position accuracy	0.05–0.3 m

As part of the DARPA GPS Guidance Package (GGP) Program, the program contractors were able to demonstrate the first, tightly-coupled INS using fiber optic gyroscopes (FOGs), a miniature GPS receiver, silicon accelerometers and 21–26 state Kalman filters [29]. This program led to the Northrop Grumman LN-251, LN-271, LN-100 and other

FOG-based INS product line, many of which are the workhorse of the industry. The cost of these navigation-grade systems has come down over the years; today, a three-axis navigation-grade inertial navigation system like LN-251 runs around $85,000. This is a significant drop in price compared to the $1M price tag in the late 1980s. Similar capabilities were first applied to the synthetic aperture radars [30, p. 370] to provide this level of accuracy over relatively long integration times. An in-depth discussion is beyond the scope of this book. For the interested reader a good technical discussion of this topic is given in [1, p. 195].

In the examples presented in this chapter it is easy to see where accurate geo-location is important and where it isn't. For example mapping the atmospheric constituents, while important, does not require centimeter accuracy. On the other hand, delivering a weapon payload often does to avoid collateral damage.

9.7   Summary

We have considered a variety of lidar problems, ranging from a very simple ranging system that has a broad application in traffic control, to a fairly sophisticated detection system for underwater objects. All the applications are based upon accurate models, but the computation of these models in some cases is a little daunting. Nevertheless, even without detailed computer computations, estimates can be made for the performance of most systems. We have also conjectured on the possible usefulness of bi-statics.

References

1. J. Shan and C. K. Toth, eds. *Topographic Laser Ranging and Scanning, Principles and Processing*. CRC Press, New York (2009).
2. V. A. Kovalev and W. E. Eichinger. *Elastic Lidar: Theory, Practice and Analysis Methods*. John Wiley, New York (2004).
3. M. I. Skolnik, ed. *Radar Handbook*. McGraw-Hill, New York (1970), p. 27.16 (1–4).
4. J. Ruze. Antenna tolerance theory – a review. *Proceedings of the IEEE*, **54**(4) (1966), pp. 633–641.
5. J. W. Goodman. Statistical properties of laser speckle patterns. *Topics in Applied Physics*, **9** (1975), 9–75. doi: 10.1007/BFb0111436.
6. V. Wulfmeyer and J. Bösenberg. Ground-based differential absorption lidar for water-vapor profiling: assessment of accuracy, resolution, and meteorological applications. *Applied Optics* 37 (1998), pp. 3825–3844.
7. E. J. McCartney. *Optics of the Atmosphere*. John Wiley and Sons, New York (1976), pp. 176–215.
8. D. N. Whiteman. Examination of the traditional Raman lidar technique. I. Evaluating the temperature-dependent lidar equations. *Applied Optics*, **42**(15) (2003), pp. 2571–2592. II. Evaluating the ratios for water vapor and aerosols. *Applied Optics*, **42**(15) (2003), pp. 2593–2608.
9. USSA (1962), U.S. Standard Atmosphere, 1962, Cat. No. NAS1.2:At6/962. GPO, Washington DC.
10. J. A. Curcio and C. L. Knestric. Correlation of the atmospheric transmission and backscattering. *Journal of the Optical Society of America*, **48** (1958), pp. 686–689.

11. W. Viezee, E. E. Uthe and R. T. H. Collis. Lidar observations of airfield approach conditions: an exploratory study. *Journal of Applied Meteorology*, **8** (1969), pp. 274–283.

12. J. D. Klett. Lidar inversion with variable backscatter/extinction ratios, *Applied Optics*, **24** (1985), pp. 1638–1643.

13. W. E. Feltz, W. L. Smith, R. O. Knuteson *et al.* Meteorological applications of temperature and water vapor retrievals from the ground-based Atmospheric Emitter Radiance Interferometer (AERI). *Journal of Applied Meteorology*, **37** (1998), pp. 857–875.

14. A. Behrendt, T. Nakamura, M. Onishi, R. Baumgart and T. Tsuda. Combined Raman lidar for the measurement of atmospheric temperature, water vapor, particle extinction coefficient, and particle backscatter coefficient. *Applied Optics*, **41**(36) (2002), pp. 7657–7666.

15. W. Saeys, B. Lenaerts, G. Craessaerts and J. de Baerdemaeker. Estimation of the crop density of small grains using LIDAR sensors. *Biosystems Engineering*, **102**(1) (2009), pp. 22–30.

16. M. E. Hodgson, J. Jensen, G. Raber *et al.* An evaluation of lidar-derived elevation and terrain slope in leaf-off conditions. *Photogrammetric Engineering & Remote Sensing*, **71**(7) (2005), pp. 817–823.

17. B. Tatarov, D. Muller, D. H. Shin *et al.* Lidar measurements of Raman scattering at ultraviolet wavelength from mineral dust over East Asia. *Optics Express*, **19**(2) (2011), pp. 1569–1581.

18. A. J. Sedlacek III, M. D. Ray, N. S. Higdon and D. A. Richter. Short range, non-contact detection of surface contamination using Raman lidar. *Proceedings of the SPIE*, **4577**, 95–104 (2001).

19. Z. Liu, N. Sugimoto and T. Murayama. Extinction-to-backscatter ratio of Asian dust observed with high-spectral-resolution lidar and Raman lidar. *Applied Optics*, **41**(15) (2002), pp. 2760–2767.

20. J. Llorens, E. Gil, J. Llop and M. Queralto. Georeferenced LiDAR 3D vine plantation map generation. *Sensors*, **11** (2011), pp. 6237–6256. doi: 10.3390/s110606237.

21. Z. Hans, R. Tenges, S. Hallmark, R. Souleyrette and S. Pattnaik. Use of LIDAR-based elevation data for highway drainage analysis: a qualitative assessment. *Proceedings of the Mid-Continent Research Symposium*, Ames, Iowa, August 2003.

22. K. F. G. Olofson, Georg Witt and Jan B. C. Pettersson. Bistatic lidar measurements of clouds in the Nordic Arctic region. *Applied Optics*, **47** (2008), pp. 4777–4786.

23. K. Meki, K. Yamaguchi, X. Li, Y. Saito, T. D. Kawahara and A. Nomura. Range-resolved bistatic imaging lidar for the measurement of the lower atmosphere. *Optics Letters*, **21** (1996), pp. 1318–1320.

24. B. M. Welsh and C. S. Gardner. Bistatic imaging lidar technique for upper atmospheric studies. *Applied Optics*, **28** (1989), pp. 82–88.

25. L. B. Stotts. Closed form expression for pulse broadening in multiple scattering media. *Applied Optics*, **17** (1978), pp. 504–505.

26. C. S. Tan, A. Sluzek and G. Seet. Model of gated imaging in turbid media. *Optical Engineering*, **44**(11) (2005).

27. C. S. Tan, G. Seet, A. Sluzek *et al.* Scattering noise estimation of range-gated imaging system in turbid condition. *Optics Express*, **18**(20) (2010).

28. S. Karp. Advanced optical scatter experimental design. RADC, Final Technical Report, RL-TR-91–232, September 1991, pp. 38.

29. L. B. Stotts and J. Aein. Guidance technology useful for military communications. *Proceedings of MILCOM*, **89**, 15–18 October 1989.

30. J. C. Curlander and R. N. McDonough. *Synthetic Aperture Radar*. John Wiley & Sons, New York (1991).

10 Communications in the turbulence channel

10.1 Introduction

Since lasers were invented in 1964, optical communications has been investigated for both military and commercial application because of its wavelength and spectrum availability advantages over radio frequency (RF) communications. Unfortunately, only fiber optic communications (FOC) systems have achieved wide implementation since then because of their ability to maximize power transfer from point to point while also minimizing negative channel effects. Recently, free-space optical communications (FSOC) has reemerged after three decades of dormancy due to the availability of new FOC technologies to the FSOC community, such as low-cost sensitive receivers and more power-efficient laser sources. Applications of FSOC, however, have been limited to local area (short range) networking because optical systems have been unable to effectively compensate for two atmospheric phenomena; cloud obscuration and atmospheric turbulence. Making a hybrid FSOC/RF communications system will compensate for cloud obscuration by using the RF capability when "clouds get in the way". For this chapter, we will discuss how to mitigate the atmospheric turbulence for incoherent communications systems. In particular, we will discuss a new statistical link budget approach for characterizing FSOC link performance, and compare experimental results with statistical predictions. We also will comment at the end of the chapter on progress in coherent communications through turbulence. For those readers interested in science and modeling of laser propagation through atmospheric turbulence, we refer them to several excellent books on the topics for the details [1–5].

10.2 Degradation from atmospheric absorption and scattering

In addition to other optical loss mechanisms, the power at the receiver is reduced by the absorption and scattering by molecules and aerosols along the path. The transmission over a constant altitude path follows Beer's law and is given by

$$L_a = \exp\left[-c\left(\lambda\right)R\right] \tag{10.1}$$

where L_a is the atmospheric transmittance over the distance R, $c\left(\lambda\right) = a\left(\lambda\right) + b\left(\lambda\right)$ is the volume extinction coefficient (m^{-1}), $a\left(\lambda\right)$ is the volume absorption coefficient (m^{-1}) and $b\left(\lambda\right)$ is the volume scattering coefficient (m^{-1}). All are a function of wavelength.

When the altitude is not constant, we integrate this transmission over the path length. Software packages like LOWTRAN, FASCODE, MODTRAN, HITRAN and PCLNWIN can be used to estimate this transmittance as a function of wavelength. Currently, LOWTRAN is no longer used because it was a low-resolution wavelength solution. MODTRAN has broadband precision at its moderate resolution and can calculate atmospheric transmission over various slant paths. It also has many variables, e.g., visibility, time of day, weather, with which to do trade-offs. FASCODE provides high-resolution atmospheric transmission estimates, while HITRAN is a database with laser line precision.

The models the authors find most useful are the ones in *The Infrared Handbook* [6]. Specifically, we use the estimates of the attenuation coefficients in the Rural Aerosol Model and the Maritime Aerosol Model as a function of wavelength and in the variation with altitude for Moderate Volcanic conditions in that reference. For example, at a wavelength of 1.55 μm, the Rural Aerosol Model attenuation coefficient is given as 0.036 per kilometer and the Maritime Aerosol Model is 0.120 per kilometer. We recognize that the maritime attenuation is appreciably greater than the rural value, due largely to the presence of salt aerosols.

10.3   Atmospheric turbulence and its characterization

In the marine and atmospheric channels, turbulence is associated with the random velocity fluctuations of the "viscous fluid" comprising that channel. Unfortunately, these fluctuations are not laminar in nature, but rather are random sub-flows called turbulence eddies [1–5], which are generally unstable. In 1941, Kolomogorov, and simultaneously Obukhov, analyzed an isotopic velocity field to derive a set of inertial-subrange predictions for the forms of the velocity vector field and velocity structure functions that bear his name [1, p. 20]. Both researchers independently found from their analyses that these eddies can be broken up into a range of discrete eddies, or turbules; in particular, a continuum of cells for energy transfer from a macro scale, known as the outer scale of turbulence L_0, down to micro scale, known as the inner scale of turbulence ℓ_0 [3]. This is illustrated in Figure 10.1 showing the energy process. Near the surface (< 100 m), the outer scale L_0 is of the order of the height above ground level (agl) of the observation point and the inner scale ℓ_0 typically is of the order 1–10 mm. This range of turbules affects the local temperature and pressure structure, causing changes in the medium's refractive index. The twinkling of the stars at night is a manifestation of the dynamic nature of the refractive index changes induced by these cells. In this section, we will provide a general background of the mathematical description of turbulence relevant to FSOC link performance. This information will provide the basis for the characterization of laser propagation in the turbulent channel described in this chapter.[1]

---

[1] This chapter ignores the propagation in marine turbulence as particular scattering masks its effects over link distances of any real utility.

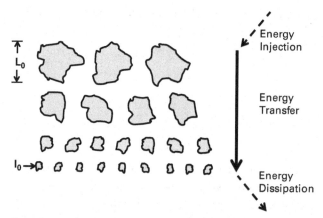

Figure 10.1 Kolmogorov cascade theory of turbulence, where L_0 denotes the outer scale of turbulence and l_0 denotes the inner scale of turbulence [2, p 60].

The index of refraction depends mostly on the local pressure and temperature and is written as

$$n\,(\mathbf{r}) \;=\; 1 + 79x10^{-6}\,[P(\mathbf{r})/T(\mathbf{r})] \;=\; 1 + n_1\,(\mathbf{r}) \tag{10.2}$$

where $\mathbf{r}$ is a vector location in space [3, p. 10].

Because turbulence is a random process, it must be analyzed with statistical quantities. The turbulent fluctuations in the channel are best described by the structure function

$$D_n\,(\mathbf{r}_1,\,\mathbf{r}_2) \;=\; \langle\,|\,n\,(\mathbf{r}_1) - n\,(\mathbf{r}_2)\,|^2\,\rangle \tag{10.3}$$

where <...> represents ensemble averaging, and n is the refractive index at vector location in the argument.

For homogeneous and isotropic turbulence, the refractive index structure function only depends on the modulus of the vector separation between the two vector locations. This means that Eq. (10.3) can be rewritten as

$$D_n\,(\mathbf{r}_1,\,\mathbf{r}_2) \;=\; D_n\,(\mathbf{r}_1,\,\mathbf{r}_2) \;=\; D_n\,(\rho) \tag{10.4}$$

from the Kolmogorov-Obukhov analyses, the refractive index structure function follows a two-thirds power law.

$$D_n\,(\rho) \;=\; \langle\,|\,n\,(\mathbf{r}_1) - n\,(\mathbf{r}_2)\,|^2\,\rangle \;=\; C_n^{\,2}\,(h)\rho^{2/3}, \quad \text{for } \ell_0 < \rho < L_0 \tag{10.5}$$

where $C_n^{\,2}(h)$ is called the refractive index structure parameter. This parameter is a function of atmospheric temperature and pressure, as well as wavelength. In general, $C_n^{\,2}$ varies strongly with height above ground; it is largest near the ground and falls off dramatically with increasing height above ground. It also follows a diurnal cycle with peaks at mid-day, near constant values at night and lowest near sunrise [3, p. 11]. Weak turbulence typically has refractive index structure parameters around 10^{-17} m$^{-2/3}$ and strong turbulence has values around 10^{-13} m$^{-2/3}$ or greater.

Another parameter of interest is the power spectrum model for the turbulence, which is the Fourier transform of the refractive index covariance function. From the two-thirds power law, the power spectral density of the refractive index fluctuations can be derived as

$$\Phi_n(\kappa) = 0.033 \, C_n^2 \, \kappa^{-\frac{11}{3}}, \quad \text{for } \kappa_0 < \kappa < \kappa_\ell \tag{10.6}$$

where

$$\kappa_0 = 2\pi/L_0 \tag{10.7}$$

and

$$\kappa_h = 2\pi/\ell_0 \tag{10.8}$$

The above spectrum is widely used in theoretical analysis, but, as noted, has a limited inertial subrange. This limitation requires other models to be used in some calculations.

One such alternative power spectrum model was proposed by Tartarski for $\kappa \gg 1/\ell_0$. In particular, he wrote

$$\Phi_n(\kappa) = 0.033 \, C_n^2 \, \kappa^{-\frac{11}{3}} \exp\left(-\kappa^2/\kappa_m^2\right), \quad \text{for } \kappa_0 \ll \kappa \tag{10.9}$$

where $\kappa_m = 5.92/\ell_0$. $\kappa < 1/L_0$. In general, the spectrum is anisotropic and its mathematical form of the power spectrum density is unknown [3, p. 13].

Another model, called a modified von Karmon spectrum, is given by

$$\Phi_n(\kappa) = 0.033 \, C_n^2 \, \frac{\exp\left(-\kappa^2/\kappa_m^2\right)}{\left(\kappa^2 + \kappa_0^2\right)^{11/6}}, \quad \text{for } 0 < \kappa < \infty \tag{10.10}$$

where $\kappa_0 = 2\pi/L_0$.

Andrews notes that none of the above models contains the rise (or "bump") at high wavenumbers near $1/\ell_0$ that is found in temperature spectral data [3, p. 13]. To alleviate this situation, he provided a modified atmospheric approximation that includes that bump: namely,

$$\Phi_n(\kappa) = 0.033 \, C_n^2 \left[1 + 1.802 \, (\kappa/\kappa_\ell) \right.$$

$$\left. - 0.254 \, (\kappa/\kappa_\ell)^{7/6} \, \frac{\exp\left(-\kappa^2/\kappa_\ell^2\right)}{\left(\kappa^2 + \kappa_0^2\right)^{11/6}} \right], \quad \text{for } 0 < \kappa < \infty \tag{10.11}$$

where $\kappa_\ell = 3.3/\ell_0$.

For vertical or slant paths, the structure parameter C_n^2 must be described in terms of an altitude profile model. There are many atmospheric models that have been developed over the years, e.g., the so-call HV 5/7, the clear-1 model, the ESO-Parnal model, the SLC-Night model, the SLC-Day model, and the AMOS-Night model [7]. The model the authors find most useful is the HV 5/7 model because of its traceability to statistical link availability. HV 5/7 stands for the Hufnagel-Valley (HV) 5/7 $C_n^2(h)$ model. Here, the term 5/7 means that for a wavelength of 0.5 μm, the value of 5 represents a Fried parameter of 5 cm and the value of 7 represents an isoplanatic angle for a receiver on the ground looking up of 7 µrad. (These terms will be discussed in the next section on the point spread function.) Mathematically, the profile is given by

$$C_n^2(h) = 0.00594 \left(w/27 \right)^2 \left(10^{-5} h \right)^{10} \exp\left[- h/1000 \right]$$

$$+ 2.7x10^{-16} \exp\left[- h/1500 \right] + A \exp\left[- h/100 \right] \quad (10.12)$$

where $A = C_n^2(0)$ is the ground-level value of the structure parameter and w is the rms wind speed given by the equation

$$w = \sqrt{ \frac{1}{15x10^3} \int_{5\times10^3}^{20\times10^3} V_{HV5/7}(h) \, dh }$$

$$= \sqrt{ \frac{1}{15x10^3} \int_{5\times10^3}^{20\times10^3} \left\{ w_s h + w_g + 30 \exp\left(- \left[\frac{h-9400}{4800} \right]^2 \right) \right\}^2 dh } \quad (10.13)$$

where $V_{HV\,5/7}(h)$ is the HV 5/7 vertical profile for wind speed, w_g is the ground wind speed and w_s is the beam slew rate associated with a satellite moving with respect to an observer on the ground [2, p. 481; 3, p. 12]. Figure 10.2 illustrates two HV $C_n^2(h)$ profiles for two values of $C_n^2(0)$ (= 1.7×10^{-14} m$^{-2/3}$ for HV 5/7) and three upper atmosphere wind pseudo-wind speeds ($w = 21$ m/s for HV 5/7). It is clear from this figure that the value of $C_n^2(0)$ affects the structure parameter below 1 km altitude and has no effect on it above that height. It also is clear that the value of the pseudo-wind speed has little effect on the structure parameter below 5 km altitude, but affects it above that height. Figure 10.3 compares multiples of the HV 5/7 model compared with annual Korean turbulence statistics. In this figure, we have 0.2, 1 and 5 × HV 5/7 model values plotted against turbulence occurrence statistics of 15, 50 and 85%, respectively (the percentages

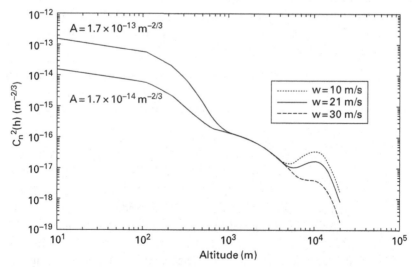

Figure 10.2 Vertical distribution of atmospheric turbulence C_n^2 for the Hufnagel-Valley model.

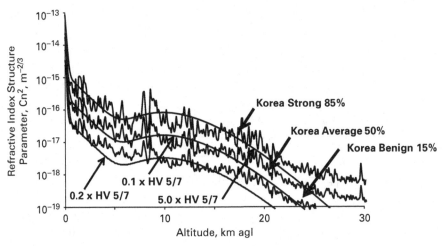

Figure 10.3 Multiples of Hufnagle-Valley (HV) 5/7 model compared to Korean turbulence.

in the legend reflect the amount of time during the year the value of C_n^2 occurred). This figure shows reasonable agreement between these models and the yearly statistics.

It has long been recognized that the advantage of the HV model over other atmospheric models is its inclusion of the rms wind speed, w, and the ground-level refractive index structure parameter, $C_n^2(0)$. Their inclusion permits variations in high-altitude wind speed and local near-ground turbulence conditions to better model real-world profiles over a large range of geographic locations. It also provides a model consistent with measurements of the Fried parameter, r_0, and the isoplanatic angle, θ_0, both of which will be discussed in the next section. However, Andrews and Phillips noted that the last exponential term in Eq. (10.12) describes the near ground turbulence conditions, which slowly decreases in $C_n^2(h)$ with altitude up to approximately 1 km [8]. This is in conflict with the $C_n^2(h)$ behavior of $h^{-4/3}$ noted by Walters and Kunkel [9] and supported by a number of other early measurements. To better represent this trend, Andrews and Phillips modified the Hufnagle-Valley to yield

$$C_n{}^2(h) = M\,[0.00594\,(w/27\,)^2\,(10^{-5}\,(h + h_s))^{10}\exp\,[-(h + h_s)/1000]$$

$$+\,2.7\mathrm{x}10^{-16}\exp\,[-(h + h_s)/1500] + C_n{}^2(0)\,(h_0/h\,)^{4/3}\,] \qquad (10.14)$$

for $h > h_0$. They refer to this model as the Hufnagle-Andrews-Phillips (HAP) model [8,10]. Comparing Eqs (10.12) and (10.14), we see that (1) the last exponential in the former equation has been replaced in the latter by a term that reflects the observed behavior by Walters and Kunkel, (2) a reference height of the ground above sea level, h_s, has been added, and (3) a scaling factor, M, also added to represent the strength of the average high-altitude background turbulence. Figure 10.4 compares the HV and HAP models for $h_s = 0$ m, $h_0 = 1$ m, $M = 1$, $w = 21$ m/s and $C_n^2(0) = 1.7 \times 10^{-14}$ m$^{-2/3}$. For the first few hundred meters, there is considerable difference between the models. However, at approximately 1 km and higher altitudes, the models are essentially the same.

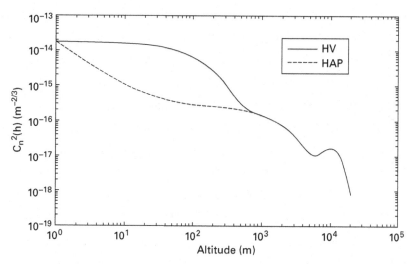

Figure 10.4 Comparison of Hufnagel-Valley (HV) and Hufnagel-Andrews-Phillips (HAP) models.

Because the power law behavior as a function of altitude changes from $h^{-4/3}$ during the day to $h^{-2/3}$ at night, it is clear that there must be a transition period between day and night. Andrews *et al.* [10] have recently developed a transition model that varies like h^{-p}, where p is dependent upon the temporal hour of the day. A temporal hour is defined as 1/12 the number of hours between sunrise and sunset. This more general model therefore makes use of the actual sunrise and sunset times, and particular time of day under which experiments are performed, to determine the value of p.

As a final note, some researchers feel that wind speed estimated by Eq. (10.13) is too high and replace it with the Bufton model, which is given by

$$
w = \sqrt{\frac{1}{15 \times 10^3} \int_{5x10^3}^{20x10^3} V_{Bufton}(h) \, dh}
$$

$$
= \sqrt{\frac{1}{15 \times 10^3} \int_{5x10^3}^{20x10^3} \left\{ w_s h + w_g + 20 \exp\left(-\left[\frac{h-12000}{6000}\right]^2\right)\right\}^2 dh} \quad (10.15)
$$

where V_{Bufton} is the Bufton vertical profile for wind speed, $w_g = 2.5$ m/s [7]. Figure 10.5 compares the HV 5/7 with the Bufton model, which clearly shows the latter's smaller wind profile.

In most applications in the turbulence channel, one will evaluate link performance over slant paths in the atmosphere rather than the vertical paths implied above. This means that the refractive index structure parameter (and the volume extinction coefficient for that matter) must be evaluated over that path. The following provides the required translation from the vertical path formulation.

Figure 10.6 depicts assumed communications geometry. Let R_e represent the radius of the Earth, h_t be height above the Earth's surface of the transmitter, h_r be height above the

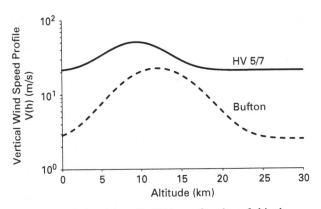

Figure 10.5 Wind speed profile V(h) as a function of altitude.

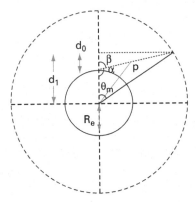

Figure 10.6 Atmospheric slant path model geometry.

Earth's surface of the receiver and h be the height above the Earths' surface of the point p. This implies that $R + h_t$ is the distance above the center of the Earth to the transmitter, $R + h_r$ is the distance above the center of the Earth to the receiver and $R + h_r$ is the distance above the center of the Earth to the point p.

From this figure, we see that

$$d_1 = (R_e + h_r)\cos \theta_m \tag{10.16}$$

which leads to

$$d_0 = (R_e + h_r)\cos \theta - (R_e + h_r) \tag{10.17}$$

and

$$c = (R_e + h_r)\sin \theta_m \tag{10.18}$$

Therefore, we can write the angles β and α as

$$\beta = \tan^{-1} \frac{(R_e + h_r)\sin \theta_m}{(R_e + h_r)\cos \theta_m - (R_e + h_t)} \tag{10.19}$$

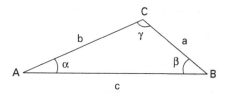

Figure 10.7 Obtuse triangle.

and

$$a = \left[\pi - \tan^{-1}\frac{(R_e + h_r)\sin\theta_m}{(R_e + h_r)\cos\theta_m - (R_e + h_t)}\right], \tag{10.20}$$

respectively. The distance between the transmitter and receiver, called the range, R, is given by

$$R = (R_e + h_r)\sin\theta_m \csc\left[\tan^{-1}\frac{(R_e + h_r)\sin\theta_m}{(R_e + h_r)\cos\theta_m - (R_e + h_t)}\right] \tag{10.21}$$

Law of sines and cosines

Figure 10.7 depicts a triangle with three sides of length a, b and c and angles α, β and γ, respectively.

The law of sines is

$$\frac{a}{\sin a} = \frac{b}{\sin\beta} = \frac{c}{\sin\gamma}$$

The law of cosines is

$$c^2 = a^2 + b^2 - 2ab\cos\gamma$$

$$b^2 = a^2 + c^2 - 2ac\cos\beta$$

and

$$a^2 = c^2 + b^2 - 2cb\cos\alpha$$

From the law of cosines, we have the following for the point p

$$(R_e + h)^2 = R_e + s^2 - 2R_e s\cos\alpha \tag{10.22}$$

for $h_t \le h \le h_r$. Recall that

$$\cos A - B = \cos A\cos B - \sin A\sin B \tag{10.23}$$

This implies that

$$(R_e + h)^2 = R_e^2 + s^2 - 2R_e s\cos\beta \tag{10.24}$$

using Eq. (10.19). Expanding the above equation allows us to write

$$h^2 + 2R_e h - [s^2 + 2Rs\cos\beta] = 0 \tag{10.25}$$

Using the positive solution of the quadratic equation, we see that

$$h = \sqrt{R_e^2 + s^2 + 2R_e s \cos\beta} - R_e \qquad (10.26)$$

Referring to Figure 10.5 again as well as the law of cosines, we have

$$\vartheta_m = \cos^{-1}\frac{(R_e + h_t)^2 + (R_e + h_r)^2 - R_e^2}{2(R_e + h_t)(R_e + h_r)} \qquad (10.27)$$

From Eq. (10.26), we know that

$$\cos\vartheta_m = \frac{(R_e + h_t)^2 + (R_e + h_r)^2 - R^2}{2(R_e + h_t)(R_e + h_r)} \qquad (10.28)$$

$$\cos\vartheta_m \approx 1 - \frac{R^2}{2R_e^2} \qquad (10.29)$$

since $R_e \gg h_t, h_r$. We also know that $R_e \gg R$, which means that

$$\vartheta_m \approx \frac{R}{R_e}. \qquad (10.30)$$

Going back to Eq. (10.19), we find that

$$\beta \approx \tan^{-1}\frac{R}{(R_e + h_r) - (R_e + h_t)} \qquad (10.31)$$

$$\approx \tan^{-1}\frac{R}{h_r - h_t} \qquad (10.32)$$

since $h_t \leq h_r$.

Example 10.1 Let

$$h_t = 0.061 \text{ m}$$

$$h_r = 0.561 \text{ m}$$

and

$$R = 10 \text{ km}.$$

Then,–

$$\beta \approx \tan^{-1}\frac{R}{h_r - h_t} \approx \tan^{-1}\frac{10 \text{ km}}{0.561 \text{ km} - 0.061 \text{ km}} \approx \tan^{-1}20 \approx 87.139 \text{ degrees}$$

and

$$\cos\beta = 0.04994.$$

Given that $R_e \gg s$, $0 \leq s \leq R$, Eq. (10.25) reduces to

$$h \approx s \cos\beta = s^* 0.04994$$

which scales the same value of the arcsine as using the ratio $(500 \text{ m} / 10\,000 \text{ m}) = 0.05$.

10.4 The point spread function created by atmospheric turbulence and other important entities

When the incoming beam traverses atmospheric turbulence before it illuminates the circular aperture, the resulting point spread function from the circular aperture will be limited by diffraction and the turbulence the beam encountered. In particular, the point spread function in this case will be a combination of expected point spread function defined by λ/D and a "halo" about this peak with a width defined by λ/r_0, where r_0 is the transverse coherence length of the atmosphere, better known as the Fried parameter.

The Fried parameter is generically defined as

$$r_0 = \left[0.42\, k^2 \sec(\zeta) \int_{Path} C_n^{\;2}(r)\, dr \right]^{-3/5} \qquad 10.33$$

where ζ is the zenith angle, $k \equiv$ wavenumber $= 2\pi/\lambda$ and $C_n^{\;2}(r)$ is the refractive index structure parameter, as noted earlier [3, p. 47]. It is the atmosphere's lateral coherence distance, over which the phase also varies no more than $\pm\pi$. It is related to the plane wave coherence radius, ρ_0, by the relation

$$r_0 = 2.1\rho_0. \qquad (10.34)$$

[3, p. 47]. At sea level, r_0 is usually 2–15 cm in daytime at visible wavelengths. In practice, r_0 is of the order of the speckle size of the turbulence at the receiver plane; this point will become important in a discussion to come.

In Chapter 2, we introduced the point spread function for rectangular and circular apertures, as well as the optical transfer function (OTF), which is the Fourier transform of the normalized point spread function, i.e., the transfer function of an incoherent imaging system. The modulus of the OTF is called the modulation transfer function (MTF). In free space, the MTF of an imaging system is the following:

$$\mathrm{MTF}_0(v) = \frac{2}{\pi} \left[\cos^{-1}(\lambda f\, v/D) - (\lambda f\, v/D) \sqrt{1 - (\lambda f\, v/D)^2} \right], v < D/\lambda f \qquad (10.35a)$$

$$= 0,\, v > D/\lambda f \qquad (10.35b)$$

where v represents spatial frequency [3, p. 48]. In a turbulent atmosphere, the MTF is given by

$$\mathrm{MTF}_0(v) = \exp\left[-(\lambda f\, v/r_0)^{5/3} \right] \qquad (10.36)$$

neglecting particulate scattering [3, p. 48]. The total MTF for an imaging system in a turbulent atmosphere is then

$$\mathrm{MTF}_{total}(v) = \mathrm{MTF}_0(v) \times \mathrm{MTF}_{atm}(v) \qquad (10.37)$$

The resulting point spread function is obtained by Fourier transforming Eq. (10.37).

For our discussion to come, we expect to have transmitters that have expanding beams and receivers with expanding fields of view, i.e., spherical waves. Under these conditions, we use a modified form of the Fried parameter to reflect the receiver and transmitter coherence lengths of spherical waves in the later part of this chapter; specifically, the receiver and transmitter coherence lengths are given by

$$r_{0R} = \left[16.71 \sec(\zeta) \int_0^L C_n^2(r) \left(r/R \right)^{5/3} dr/\lambda^2 \right]^{-3/5} \quad (10.38)$$

and

$$r_{0T} = \left[16.71 \sec(\zeta) \int_0^L C_n^2(r) \left(1 - r/R \right)^{5/3} dr/\lambda^2 \right]^{-3/5} \quad (10.39)$$

respectively. Here, L is the length of the turbulent regime (see Beland [4], Eqs. 2.135 and 2.155). In this situation, the far field patterns for the transmitter and receiver have angular extent of λ/r_{0T} and λ/r_{0R}, respectively. In vertical link geometries, there are two other parameters helpful in analyzing laser propagation or imaging through atmospheric turbulence. They are the isoplanatic angle and the Greenwood time constant.

When light traveling through the atmosphere has a changing frame of reference due to platform(s) motion, there might be the occasion when the statistics of the atmosphere change due to that motion. That is undesirable. The isoplanatic angle defines the mean-square error of 1 radian-squared (rad^2) between the paths and is given by

$$\theta_0 = \frac{\cos^{8/5}(\zeta)}{\left[2.91 \, k^2 \int_{h_0}^H C_n^2(h) \, (h - h_0)^{5/3} \, dh \right]^{3/5}} \quad (10.40)$$

where h is the height above the ground [3, p. 51]. At $\lambda = 0.5 \, \mu m$, θ_0 is roughly 7–10 μrad for near-vertical paths from Earth to space.

The Greenwood frequency, f_g, is the characteristic frequency of atmospheric statistical change. Big turbules move around slowly and small turbules move around much more frequently, but on the average the entirety changes at this frequency. The time interval over which turbulence remains essentially unchanged is called the Greenwood time constant, and it is the inverse of the Greenwood frequency. Mathematically, we relate both by the equation

$$\tau_0 = \frac{\cos^{3/5}(\zeta)}{\left[2.91 \, k^2 \int_{h_0}^H C_n^2(h) \, V^{5/3}(h) \, dh \right]^{3/5}} = 1/f_g \quad (10.41)$$

[3, p. 47]. This time constant is typically of the order of milliseconds and the Greenwood frequency is in the 100s of hertz. Let us now move on to characterizing the irradiance.

Recall from example 2.7 that we introduced the Strehl ratio as a good measure of the quality of the incoming irradiance to an optical system [11, p. 462]. The concept was introduced by Herr Professor K. Strehl in 1902 [12] and amplified by V. N. Mahajan, in the 1980s [13,14]. Andrews and Phillips extended its use to the turbulent atmosphere

[2, pp. 407–409]. For all turbulence conditions, they stated that the Strehl ratio is defined by the integral

$$\text{SR} = \frac{16}{\pi} \int_0^1 u \left[\cos^{-1} u - u \sqrt{1 - u^2} \right] \exp \left[-3.44 \left(u \left(\frac{D}{r_0} \right)^{5/3} \right) du \right] \quad (10.42)$$

which can be closely approximated by the expression

$$\text{SR} \approx \frac{1}{\left[1 + (D/r_0)^{5/3} \right]^{6/5}} \quad (10.43)$$

where D, as before, is the diameter of the optical systems and r_0 is the Fried parameter [3, p. 50]. Figure 10.8 compares Eqs. (10.42) and (10.43). It is clear from this figure that the approximation given by the latter equation closely matches the exact solution.

One parameter not often talked about, but which is very useful in systems analysis, is the Rytov number. The Rytov number is the log amplitude variance and is an indicator of the strength of turbulence along the integrated path (alternately, the variance of the log-intensity is known as the Rytov variance). It comes from one of the most widely used approximations for solving the wave equation, the Rytov approximation [6, pp. (6–17)–(6–21)]. The solution of the wave equation in this case presents the log amplitude χ and wave front phase φ as weighted linear sums of contributions from all random index fluctuations in the semi-infinite volume between the source and receiver. (Two other popular approximations, the near field and Born approximations, are special cases of the Rytov approximation.) For plane wave propagation, the Rytov number is given by

$$\sigma_\chi^2 = \left[0.56 \, k^{7/6} \int_0^L C_n^2(r) (L - r)^{5/6} \, dr \right]^{-3/5} \quad (10.44)$$

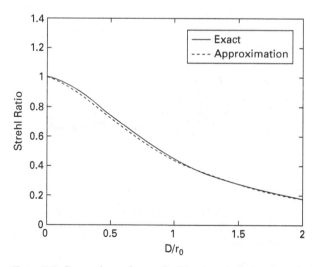

Figure 10.8 Comparison of exact Strehl ratio calculation for turbulence with Andrews and Phillips' approximate equation.

and for spherical wave propagation, it is given by

$$\sigma_\chi{}^2 = \left[0.56\, k^{7/6} \int_0^L C_n{}^2\, (r/L)\, z^{5/6}\, (L-r)^{5/6}\, dr \right]^{-3/5} \tag{10.45}$$

where L is the length of the propagation path. These equations assume that

$$\ell_0{}^2 = \; <<\lambda L << L_0{}^2,$$

with ℓ_0 and L_0 being the inner and outer scales of the turbulence [3, p. 59]. These equations predict that the Rytov number increases without limit as $C_n{}^2$, or the path length increases. In reality, this is not true and there is a limit. This can be seen in Figure 10.9. This figure depicts observed scintillation versus predicted scintillation from Rytov theory [6, pp. (6–17)–(6–21)]. It is clear from this figure that the scintillation peaks when the Rytov number reaches a value of 0.3. When the square root of the Rytov number is around 0.3 or less, little scintillation is to be expected. Above 0.3, scintillation (fading) is likely. When it exceeds 1.0, we are in the saturation regime where wave optics simulations often fail to converge and the high scintillation actually goes down slightly for increasing Rytov number. Physically, we can see what is happening from Figure 10.10. Referring to 10.10(a), when turbulence is mild $[\sigma_\chi \text{ (Rytov)} < 0.3]$, the signal will experience wavefront tilt/tip from larger turbules (large r_0), which are micro-radian deflections causing minor scintillation. Under higher turbulence condition $[0.3 < \sigma_\chi \text{ (Rytov)} < 1]$, the signal beam begins to propagate through numerous small turbules (small r_0) which has the possibility of one, or more, of them deflecting their portion of the beam into the same array element. This is shown in Figure 10.10(b). The scintillation increases, but still is in the linear regime. When we have strong to severe turbulence, the situation in Figure 10.10 is exacerbated and many small beams are mixed over the receiver plane, causing the asymptotically decreasing

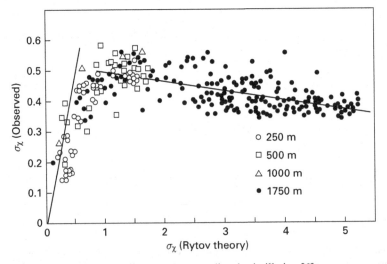

Figure 10.9 Observed scintillation versus predicted scintillation [6].

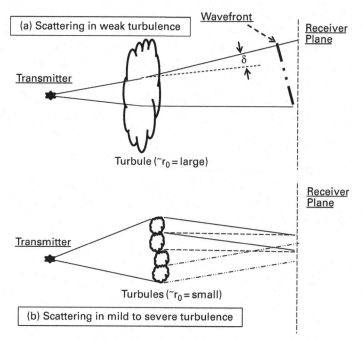

Figure 10.10 Simplistic example of turbulent scatter.

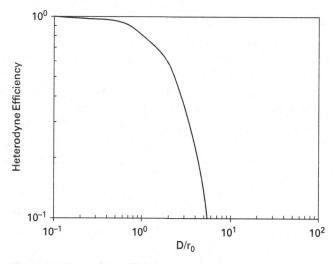

Figure 10.11 Heterodyne efficiency as a function of (D/r_0)

scintillation shown for σ_χ (Rytov) > 1. This is where 2π-ambiguity and other non-linear effects occur, the saturation regime.

Another illustration of these three regimes is shown in Figure 10.11, which shows the heterodyne efficiency as a function of (D/r_0) originally developed by Fried [15].

Referring to this figure, when the speckle is large compared to the receiver aperture diameter D, the signal beam is not distorted much and 100% heterodyne efficiency can be achieved. As r_0 exceeds D (a few speckle patches fill the aperture, or $1 < D/r_0 < 10$), the heterodyne efficiency begins to decrease. The reason is that although the amplitude errors will begin to be mitigated by aperture averaging of the speckle under this situation, the phase errors grow, causing the beating between the two beams to be degraded as the turbulence moves out of the weak regime. When numerous speckles fills the receiver aperture, $D/r_0 > 10$, the phase errors grow to the point that signal beam becomes essentially incoherent and the heterodyne efficiency goes to zero. Again, this last situation is an aspect of the saturation regime.

For the last few decades, operations and validated prediction in the saturation have had mixed success. For a laser system to have high reliability, it must be able to mitigate the effects of the saturation regime.

10.5 Adaptive optics

Turbulence in the medium to saturation regime creates beam wander, spatial and angular spread, scintillation and other negative effects on the signal beam. To mitigate these effects, adaptive optics and other non-linear techniques like four-wave mixing were invented. They also have mixed results for the last few decades. We will review the former in regard to correcting those effects in incoherent communications systems, and touch on the latter at the end of the chapter for coherent communications systems, showing how things have improved today.

Adaptive optical (AO) systems can improve optical system performance, but have limitations imposed by the size of their apertures, the number and bandwidth of their phase compensating actuators, and the distance over which they must operate. The lowest-order correction, tip/tilt, of the outgoing or incoming beam can compensate for beam wander, wherever it occurs, except for speed of light or actuator limitations. Higher-order aberrations, such as focus, astigmatism and coma, can be corrected by higher-order AO corrections, but the range of influence again is limited by receiver aperture, number of AO actuators and system bandwidth. So what is an AO system? There are many excellent texts on this subject by Robert K. Tyson, Michael C. Roggemann and Byron Welsh, and John W. Hardy [16–21]. We will not duplicate their detail here, but leave it to the interested reader to peruse. However, we will summarize some key points from these books and selected papers to benchmark the reader in AO fundamentals.

A conventional, or linear, AO system layout for imaging or laser propagation compensation is depicted in Figure 10.12. It consists of three major subsystems: (1) a wave front sensor (WFS) that detects the distorted optical signal entering the telescope; (2) an active, or deformable, mirror to correct the optically distorted signal; and (3) a control computer to monitor and decode the sensor information for the active mirror. AO theory assumes the optical signal is made up of amplitude A and phase component φ. The associated electric field then can be described as

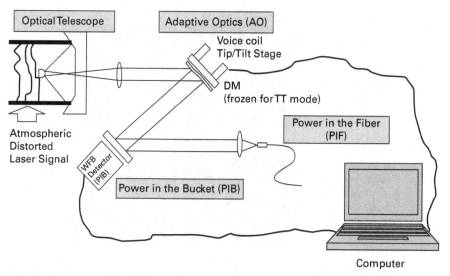

Figure 10.12 Simplified adaptive optical system layout.

$$u = A \exp\left(-i\phi\right) \tag{10.46}$$

The active/deformable mirror uses the principle of phase conjugation to correct the phase fronts of the incoming signal. In other words, it reverses the phase distortions, thereby making the spherical or planar wave input close to its original state in the absence of atmospheric effects. The reversal of the phase distortion, being in the exponent of the electric field vector shown in Eq. (10.46), means changing the sign of the term in the exponential argument. This reversal is known as "phase conjugation". What is the result of this process to the incoming degraded signal from turbulence interaction? Before we answer that, we need two key definitions. In the literature, the power at the receiver aperture is called the power in the bucket (**PIB**) and the power at the detector/ fiber is called the power in the fiber (**PIF**).

    Figure 10.13 shows a snapshot in time of the intensity in the focal plane of a curvature AO system located at one end of a 10 km FSOC link running from Saratoga, California, to Campbell, California, on 27 August 2008 [22]. Either a single or multi-mode fiber, or a detector, would be located in the center of the pictures to receive the incoming energy. When both AO systems are "off" (lower left), the received signal is off-center, very blurred and speckled. This implies radiance at the aperture entrance is scintillating and has large angular spread, plus a tilt/tip across the aperture. The PIF will suffer significantly in this situation. When the transmitter AO system is "on" and the receiver AO system is "off" (upper left), the tilt is removed, but there is still significant angular spread and scintillation of the incoming beam incident on the receiver aperture. This creates good PIB, but again poor PIF. When the transmitter AO system is "off" and the receiver AO system is "on" (lower right), the received radiance is essentially centered and much more focused. Hence, the tilt still is removed and the angular spread is reduced. To first order, it appears that both the PIB and PIF improve. Unfortunately, the signal still fluctuates a lot like in the previous

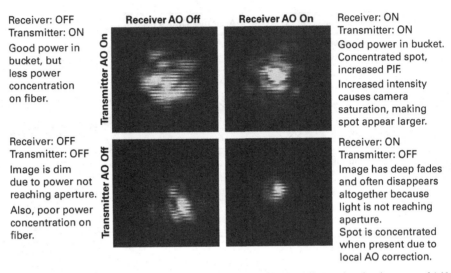

Receiver: OFF
Transmitter: ON

Good power in bucket, but less power concentration on fiber.

Receiver AO Off   **Receiver AO On**

Transmitter AO On

Receiver: ON
Transmitter: ON

Good power in bucket. Concentrated spot, increased PIF.

Increased intensity causes camera saturation, making spot appear larger.

Receiver: OFF
Transmitter: OFF

Image is dim due to power not reaching aperture.

Also, poor power concentration on fiber.

Transmitter AO Off

Receiver: ON
Transmitter: OFF

Image has deep fades and often disappears altogether because light is not reaching aperture.

Spot is concentrated when present due to local AO correction.

Figure 10.13 Focal Plane Image of AO-Compensated Received Radiance signals taken around 1600 on 23 July 2008 – Example 1.

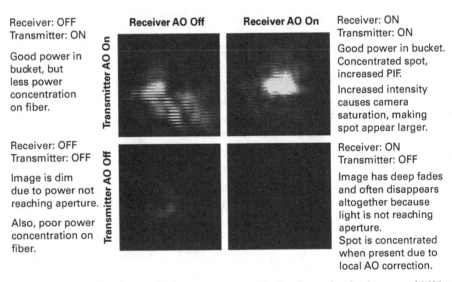

Receiver: OFF
Transmitter: ON

Good power in bucket, but less power concentration on fiber.

Receiver AO Off   **Receiver AO On**

Transmitter AO On

Receiver: ON
Transmitter: ON

Good power in bucket. Concentrated spot, increased PIF.

Increased intensity causes camera saturation, making spot appear larger.

Receiver: OFF
Transmitter: OFF

Image is dim due to power not reaching aperture.

Also, poor power concentration on fiber.

Transmitter AO Off

Receiver: ON
Transmitter: OFF

Image has deep fades and often disappears altogether because light is not reaching aperture.

Spot is concentrated when present due to local AO correction.

Figure 10.14 Focal plane image of AO-compensated received radiance signals taken around 1600 on 23 July 2008 – Example 2.

case (we will see this in the next figure.) When both AO systems are "on" (upper right), we have a reasonably good reconstructed Airy disk in the focal plane. (In the upper right picture, the center intensity has saturated the detector.) This picture implies that better coupling will occur into the detector/fiber and the PIF improves. In other words, the performance of this 10 km link is improved when both AO systems are "on".

Figure 10.14 shows a different snapshot in time of the intensity in the focal plane of a curvature AO system from the same experimental data sequence as Figure 10.13.

Although this figure exhibits the same qualitative characteristics as the previous figure for the various AO system states, the scintillation has significantly reduced the intensity in the center of each figure, except for the case when both AO systems are "on" in the upper right. Both the PIB and PIF suffer except in the latter case. Again, the performance of this 10 km link is improved when both AO systems are "on".

The optical phase can be modeled as a two-dimensional surface over the aperture. The deviation from the reference surface, e.g., flat, is the wave front error, which we sense with the WFS. In AO theory, a very useful infinite series representation of the wave front is the Zernike polynomial series. In other words, wave front aberrations are often expressed as the superposition of Zernike polynomials. Radial (index n) and azimuthal (index m) polynomials are preceded by Zernike coefficients A_{nm} and B_{nm} that completely describe the wave front up to the order specified by the largest n or m [17, p. 8]. Mathematically, we write

$$\Phi(r,\theta) = A_{00} + (1/\sqrt{2}) \sum_{n=2}^{\infty} A_{n0} R^0{}_n(r/R)$$

$$+ \sum_{n=2}^{\infty} \sum_{m=2}^{\infty} [A_{nm} \cos(m\theta) + B_{nm} \sin(m\theta)] R^m{}_n(r/R) \tag{10.47}$$

where the azimuthal polynomials are sines and cosines of multiple angles, m. The radial polynomial is given by $(n-m)/2$

$$R^m{}_n(r/R) = \sum_{s=0}^{\frac{n-m}{2}} (-1)^s (n-s)! (r/R)^{n-2s} / [s!((n+m)/2-s)!((n-m)/2)-s!] \tag{10.48}$$

[17, p. 8]. This series is especially useful in characterizing AO performance because the polynomials are orthogonal over a circle of radius R, common to most optical system designs [17, p. 8]. Figure 10.15 illustrates the first 10 Zernike polynomials. The AO system used to generate Figures 10.13 and 10.14 was a 35-Zernke-component AO system [22]. In general, tilt is what is corrected in the tip-tilt segment of a beam control system; focus is often handled separately; and deformable mirrors correct for the higher-order aberrations [23].

For the past two decades, many AO have been built and tested, primarily for astronomical purposes. The Shack-Hartmann, curvature and self-referencing interferometric AO systems are three popular examples. Andrews and Phillips did a first-order qualitative comparison of these three techniques [24] and it is summarized below.

A Shack-Hartmann AO system uses an array of lenses (called lenslets) of the same focal length that is focused onto a photon sensor array. The local tilt of the wavefront across each lens then is calculated from the position of the focal spot on the sensor. Any phase aberration can be approximated to a set of discrete tilts. By sampling an array of lenslets, all of these tilts can be measured and the whole wavefront approximated. Since only tilts are measured the Shack-Hartmann cannot detect discontinuous steps in the wavefront, and is useful only with smoothly varying field variations. It performs well in weak turbulence when $D/r_0 < \frac{1}{4}$, but not so in strong turbulence (spherical wave

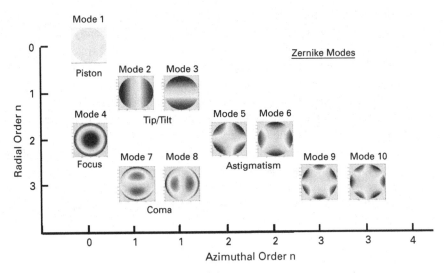

Figure 10.15 The first ten Zernike polynomials.

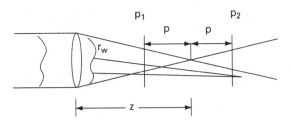

Figure 10.16 Curvature optics sensing geometry.

Rytov number > 0.2). This is because of branch point ambiguity (branch points correspond to regions of low intensity) when under strong turbulence conditions (Rytov number > 0.2).

A curvature AO system measures the local wave front curvature (second derivative) rather than the slope. Referring to Figure 10.16, it basically compares the intensity of the wave between two planes (p_1 and p_2) equidistant in front of and behind the receiver lens's focal plane at distance z. The difference in intensity, ΔI, in the two planes is proportional to the local curvature C_w of wave front, i.e.,

$$\Delta I\,(x,\,y) \;=\; \frac{2\,z^2\,C_w(x,\,y)}{p} \tag{10.49}$$

This type of AO system uses this curvature estimate to correct the incoming field. Characteristics of its first five Zernicke modes are:

- tip/tilt (x, y): displace the image (no shape change)
- focus: expand or contract the image (increase/decrease brightness)
- astigmatic (x, y): expand image in one direction and contract it in orthogonal direction.

Its advantages over the Shack-Hartmann approach are:

- eliminates the 2π phase ambiguity of Shack-Hartmann
- improves the Strehl ratio [(mean intensity)/(free-space intensity)]
- gets more light into the fiber at receiver.

A self-referencing interferometer (SRI) generates a reference wave by coupling a portion of incident wave into single-mode fiber and then creating four different wavefronts with phase shifts of 0, $\pi/2$, π, and $3\pi/2$. These phase shifts are derived from optical plates of birefringent material. An estimated phase over the wave front in the x,y plane is derived from the following arctangent relationship:

$$\arctan\left[\frac{I(3\pi/_2, x, y) - I(\pi/2, x, y)}{I(0, x, y) - I(\pi, x, y)}\right] \tag{10.50}$$

The characteristics of its first five Zernike modes and its advantages over the Shack-Hartmann approach essentially are the same as for the curvature optics.

To summarize, Andrews and Phillips made the following general comments about the three approaches [24]:

- under strong turbulence conditions (Rytov number > 0.2), curvature and SRI AO systems perform better than the Shack-Hartmann sensor because of branch point ambiguity;
- when $D/r_0 > \frac{1}{4}$, the performance of all systems begins to decline;
- smaller sub-apertures will help system performance under strong scintillation;
- assuming the Strehl ratio is the performance metric of choice:
    the Shack-Hartmann sensor performs well in weak turbulence and $D/r_0 < \frac{1}{4}$, but not
        so in strong turbulence (spherical wave Rytov number > 0.2);
    the Strehl ratio is better for both curvature sensor and SRI in strong turbulence, but
        with $D/r_0 < \frac{1}{4}$.

An estimate of the magnitude of the AO-based link performance can be found in the section titled "Random Aperture Problem" in Rayleigh's *Scientific Papers* [25]. In this section, Lord Rayleigh described the diffraction effects from a random array of 100 circular apertures, which to first order acts like a very high-order AO system. Figure 10.17 illustrates the resulting diffraction pattern [20]. Paralleling his discussion, the resulting intensity from tip/tilt alone will be of the order of N, the number of speckles at the entrance to the telescope. The resulting intensity from higher-order compensation will be of the order of N^2, the square of the number of speckles at entrance to telescope. The most improvement that higher-order AO can offer over tip/tilt-only is thus N^2/N or N. For most AO systems employed today, the possible compensation will be of the order of 4.8–15.4 dB ($N = 3$ to $N = 15$), which bounds possible compensation somewhere between 8 and 30 dB [26].

When a beam, or plane wave, propagates through weak turbulence, the first-order effect is tip/tilt, or jitter. Both describe the same phenomena, but from different viewpoints. For

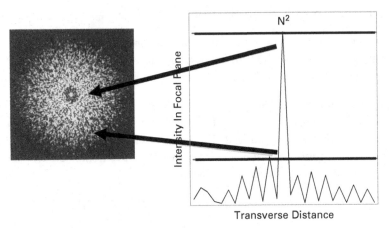

Figure 10.17 Contribution of higher order adaptive Optics [19].

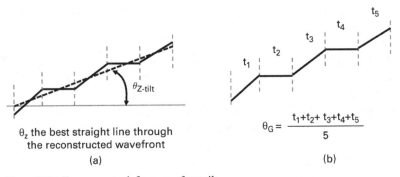

Figure 10.18 Two ways to define wavefront tilt.

example, Andrews defines the angle-of-arrival tilt variance, i.e., residual tilt at the aperture, as

$$\sigma_T^2 = \frac{6.08}{L^2 D^{1/3}} \int_0^L C_n^2(r) \, r^2 \, dr = 0.36 \left(\frac{D}{r_0}\right)^{5/3} \left(\frac{\lambda}{D}\right)^2 \tag{10.51}$$

and the tilt jitter variance for a two-axis Zernike tilt as

$$\sigma_z^2 = \frac{6.08}{D^{1/3}} \int_0^L C_n^2(r) \, dr = 0.36 \left(\frac{D}{r_0}\right)^{5/3} \left(\frac{\lambda}{D}\right)^2 \tag{10.52}$$

which yield the same result for both [3, p. 55]. Whatever you call it, dynamic tilt effects will occur even if the tracking system is doing a perfect job [3, p. 55].

When dealing with optical systems with adaptive optics or tracking, reference [27] points out that wavefront tilt at the aperture can be defined in two ways for those approaches: the two approaches are shown in Figure 10.18. Figure 10.18(a) illustrates the tilt obtained from fitting a Zernike polynomial to the wavefront. Figure 10.18(b) defines the average tilt over the aperture. The tilt variance for these two approaches is given by

$$\sigma_{\theta Z}^2 = 0.182 \, (D/r_0)^{5/3} \, (\lambda/D)^2 \tag{10.53}$$

and

$$\sigma_{\theta G}^2 = 0.170 \, (D/r_0)^{5/3} \, (\lambda/D)^2 \tag{10.54}$$

respectively. Comparing these two equations to those above for tilt, the functional dependence is exactly the same, but the multiplicative constants are approximately a factor of 2 lower. The equation for G-tilt is used for centroid calculations and the equation for z-tilt is used in adaptive optics [27].

Example 10.2. The authors of reference [27] state that there is no wavelength dependence in tilt, which is sometimes a confusion point based on the form of the above equations. Assuming constant turbulence, spherical wave propagation, they show that for G-tilt

$$\sigma_{\theta G}^2 = 0.170 \, (D/r_0)^{5/3} \, (\lambda/D)^2$$

$$= 0.170 \left(\frac{D}{(3 C_n^2 L k^2)^{-3/5}} \right)^{5/3} (\lambda/D)^2$$

$$= 0.170 \left(D^{5/3} (3 C_n^2 L \left(\frac{2\pi}{\lambda} \right)^2 \right)^2 \left(\frac{\lambda}{D} \right)^2$$

$$= 0.170(3) \, (2\pi)^2 D^{-1/3} C_n^2 L$$

[26]. Clearly, this proves their point. Although the previous equations permit using the common ratios of D/r_0 and λ/D, it can lead to an erroneous conclusion [27].

Figures 10.13 and 10.14 depict the qualitative performance of a typical curvature AO system, which probably would be similar to that of other AO system approaches. One of the key research topics for the past few decades is what kind of quantitative improvement one can expect from an AO system. In most cases, the improved performance with an AO system is dependent on the number of channels in the adaptive optics system, the diameter of the system's entrance aperture and the atmospheric conditions represented by the Fried parameter. In the next few paragraphs and the following section, we will review a selected set of developments that follow those dependences and compare their predictions with experimental data. Following this discussion, the next two sections describe alternative characterizations and approaches.

Roddier defined a performance criterion that used only the receiver diameter, the number of AO actuators and the Fried parameter in characterizing his figure of merit [28]. Specifically, he used the Strehl ratio to define the AO gain to be given by

$$G(D/r_0) = (D/r_0)^2 \exp(-0.3 \, (D/r_0)^{5/3} \, N^{-5/6}) \tag{10.55}$$

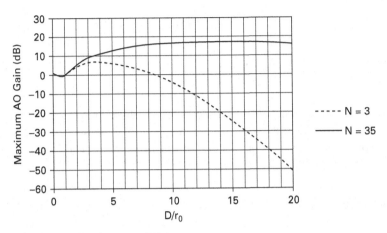

Figure 10.19 AO gain versus (D/r$_0$).

where

$$G_{\max} = 1.6\,N \tag{10.56}$$

for

$$(D/r_0) = 2.3\sqrt{N} \tag{10.57}$$

In Eqs. (10.55) to (10.57), N is the number of actuators in the AO system. Figure 10.19 illustrates the AO Gain, using Eq. (10.55), as a function of D/r_0. For a tip/tilt AO system, he projects a maximum AO gain of ~7 dB and for a 35 Zernike-component AO system, a maximum and asymptotic gain of ~17.5 dB, which are close to the gains estimated above. However, for operation in the saturation regimes, D/r_0 is less than 4 and the maximum gains are less than 10 dB.

Table 10.1 compares predictions of Roddier's equation with AO gain derived from measured Strehl ratio data taken on 27 August 2008 over a 10 km link at Campbell, California [29]. This data set contains AO performance measurements using tip/tilt only and with 35 Zernike component configurations. Fried parameter measurements were taken at the receiver site in Campbell. This table shows that the theory overestimates the measured data by 1–2 dB for data taken during the morning and evening (D/r_0 around 2 or less), but this prediction error becomes 2–3 dB different for data taken during the hotter part of the day ($D/r_0 > 3$). This estimate error follows the argument above that as D/r_0 is less than 4 and the receiver aperture is a few Fried parameters, the effectiveness of the system to mitigate turbulent effects decreases, reflecting the same type of degradation as seen for the heterodyne efficiency. Table 10.1 also shows gains less than 10 dB, as discussed in the last paragraph.

Let us look at an alternate approach to quantifying AO system performance. In his 1976 paper, Noll looked at the problem a little differently. He described the residual fitting error of an AO system as

$$\sigma_{fit}^2 = 0.29\,N^{-\sqrt{3}/2}\,(D/r_0)^{5/3} \tag{10.58}$$

Table 10.1 Comparison of Roddier's AO gain with measure data

27-August-08 Time	D/r_o	TIP/TILT (dB)		Full AO (dB)	
		Roddier $n = 3$	Strehl ratio improvement	Roddier $n = 35$	Strehl ratio improvement
09:00	2.1	4.65	3.36	6.21	5.77
13:00	3.2	6.48	3.25	9.64	7.09
14:00	4.25	6.75	4.47	11.82	8.06
18:00	1.8	3.72	3.45	4.93	4.22

[31]. Tyson points out that although Noll states that N refers to the number of Zernike modes compensated by the AO systems, it also can refer to the mean number of channels of compensation for $N > 7$ [17, p. 16]. One can use the above to yield the following equation for the Strehl ratio:

$$SR_{TT-w} \approx \frac{1}{\left[1 + 0.29 N^{-\sqrt{3}/2} (D/r_0)^{5/3} \right]} \qquad (10.59)$$

for weak turbulence and

$$SR_{TT-s} \approx \frac{1}{\left[1 + 0.29 N^{-\sqrt{3}/2} (D/r_0)^{5/3} \right]^{6/5}} \qquad (10.60)$$

for strong turbulence.

Table 10.2 compares predictions of the Noll equations for strong and weak turbulence with the measured Strehl ratios again taken on 27 August 2008 [29]. It is clear from this table that Noll's Strehl ratio for TT is 2–5 dB higher than the measured data, with more error as D/r_0 gets larger. This overestimate is about the same for the full AO comparison. An interesting observation from the results of Noll's equations is that they predict almost perfect correction for $N = 35$, which the measured data does not support.

Following the logic of Noll, Andrews has developed different Strehl ratio equations for tip/tilt and 35-Zernike-component AO systems [32]: specifically, he developed the following expressions for the tip/tilt Strehl ratio using point spread function given in Eq. (33) and the filter function given in Eq. (32), both from reference [2, p. 621].

$$SR_{TT-w} \approx \frac{1}{\left[1 + 0.28 (D/r_0)^{5/3} \right]} \qquad (10.61)$$

for weak turbulence and

$$SR_{TT-s} \approx \frac{1}{\left[1 + 0.28 (D/r_0)^{5/3} \right]^{6/5}} \qquad (10.62)$$

Table 10.2 Comparison of Noll's Strehl ratio with measure data

27-Aug-08		Tip/tilt (dB)			Full AO		
Time	D/r_0	Aoptix measured Strehl ratio $n = 3$	Noll tip/tilt Strehl ratio Weak	Noll full AO Strehl ratio Strong	Aoptix measured Strehl ratio $n = 35$	Noll full AO Strehl ratio Weak	Noll full AO Strehl ratio Strong
09:00	2.1	−4.09	−1.43822645	−1.72587174	−1.67	−0.197353661	−0.236824393
13:00	3.2	−7.21	−2.53380569	−3.04056683	−3.37	−0.389419979	−0.467303974
14:00	4.25	−8.54	−3.56258654	−4.27510385	−4.95	−0.609012956	−0.730815547
18:00	1.8	−1.31	−2.41990144	−2.90388172	−2.08	−0.153418853	−0.184102623

for strong turbulence. From the general Noll formula, the phase variance for N-35 is roughly

$$0.014\left(D/r_0\right)^{5/3}$$

which led Andrews to propose the full AO Strehl ratios to be

$$SR_{35ZM-N} \approx \frac{1}{\left[1 + 0.52\left(D/r_0\right)^{5/3}\right]} \tag{10.63}$$

for weak turbulence and

$$SR_{35ZM-N} \approx \frac{1}{\left[1 + 0.52\left(D/r_0\right)^{5/3}\right]^{6/5}} \tag{10.64}$$

for strong turbulence.

Table 10.3 compares predictions of the Andrews equations for strong and weak turbulence with the measured data again taken on 27 August 2008 [29]. This table shows that the theory for tip/tilt underestimates the measured data by 1–2 dB for data taken during the morning and evening, but increases by 2–3 dB for the data taken during the hotter part of the day. However, for full AO, the table shows that the measured data and theory for tip/tilt differ by 3–5 dB for the data taken during the morning, evening, and the hotter part of the day. Andrews and Phillips have reported similar, or better, results between theory and measurement using other link data taken by AOptix and Johns Hopkins University Applied Physics Laboratory (JHU APL) [24]. These good comparisons are in strong contrast to the comparisons using the previous two theories.

Figures 10.20 and 10.21 show a comparison of measurement and their theory for PIB and PIF using Eqs. (10.63) and (10.64) and the HAP atmospheric model. The measured data were derived from a 10 Gbps optical link running from Freemont, California, to Hollister, California, on 8 June 8 2011 from 1 to 2 p.m. Figure 10.22 depicts the terrain profile of the 17.4 km link. Figure 10.23 shows the refractive index structure parameter as a function of time for 7–9 June 2011. Regardless of the time of day that testing took place, Andrews and Philips reported that the weighted path-average C_n^2 values over the path averaged around 1.9–2.9×10^{-15} m$^{-2/3}$ all three days, derived from their three-aperture scintillometer system (TASS) [24]. Based on the weighted path-average C_n^2 values measured by TASS, the parameters M and $C_n^2(h_0)$ for the HAP model were determined; this profile model was then used to calculate the Fried parameter at both ends of the path and the mean PIB and PIF as well. The Fried parameter at the Hollister Airport end of the path ranged from 1.5 to 2.5 cm during the middle part of the day and increased up to 3.0–4.5 cm in the late afternoon. At the Fremont end of the path, the Fried parameter values were roughly 1–2 cm larger.

It is clear from Figure 10.20(a) that their theory's estimates of PIB more closely follows the average PIB measurements than implied by the comparisons in Table 10.3. Figure 10.20(b) shows their average PIF estimate is more in line with the difference noted in the above table, but in the opposite direction, i.e., overestimates the average PIF. On the other hand, Figure 10.24 compares Strehl ratio measurements and their theory for the entire day at Hollister using data and Fried parameter estimates using the

Table 10.3 Comparison of Andrews Strehl ratio with measure data

27-Aug-08		Tip/tilt (dB)			Full AO		
		Aoptix measured Strehl ratio	Noll tip/tilt Strehl ratio	Noll full AO Strehl ratio	Aoptix measured Strehl ratio	Noll full AO Strehl ratio	Noll full AO Strehl ratio
Time	D/r_0	$n = 3$	Weak	Strong	$n = 35$	Weak	Strong
09:00	2.1	-4.09	-2.93197028	-3.51836433	-1.67	-4.45721281	-5.348655372
13:00	3.2	-7.21	-4.63187711	-5.63025253	-3.37	-6.640241925	-7.96829031
14:00	4.25	-8.54	-6.15138296	-7.38165955	-4.95	-8.324152901	-9.98983481
18:00	1.8	-1.31	-2.41990144	-2.90388172	-2.08	-3.774928812	-4.529914575

(a)

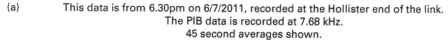

This data is from 6.30pm on 6/7/2011, recorded at the Hollister end of the link.
The PIB data is recorded at 7.68 kHz.
45 second averages shown.

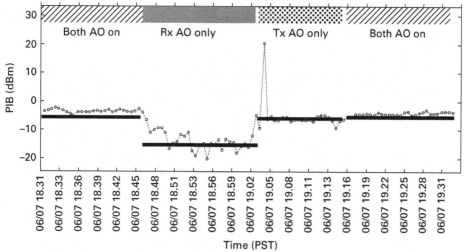

(b)

This data is from 1.00pm on 6/8/2011, recorded at the Fremont end of the link.
The PIF data is recorded at 20 kHz.
45 second averages shown.

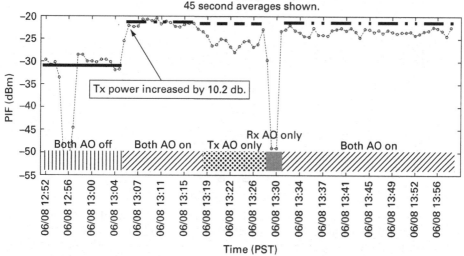

Figure 10.20 (a) Mean PIF data collected at 6:30-7:30 pm on June 7, 2011 at the Hollister Airport end of the 17-km link and (b) mean PIF data during the same time. PIB data is recorded at 7.68 kHz with 45 s averages whereas PIF data is recorded at 20 kHz with 45 s averages. The slanted-line areas represent times when the AO was turned on at both ends of the link. The dotted area represent times with Tx AO on and Rx AO off, and the black box represents times with Rx AO on and Tx AO off. The vertical line box represent the average theoretical values during these particular testing times.

HAP model independently developed by Juan Juarez, JHU APL. Here the Strehl ratio is determined from the ratio of measured average PIF to average PIB, minus optical receiver losses. In this case, the theory overestimates the Strehl ratio, but is again within a few dB of the measured values. Given the statistical nature of the channel, these are

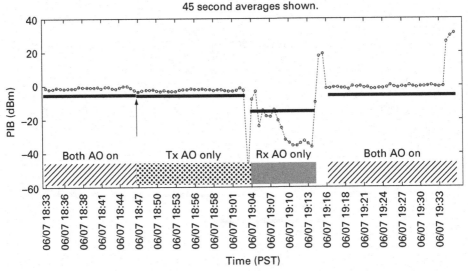

(a) This data is from 6.30pm on 6/7/2011, recorded at the Fremont end of the link. The PIB data is recorded at 7.68 kHz. 45 second averages shown.

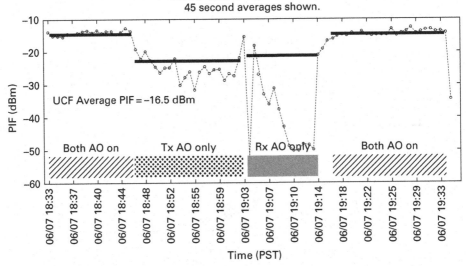

(b) This data is from 6.30pm on 6/7/2011, recorded at the Fremont end of the link. The PIF data is recorded at 20 kHz. 45 second averages shown.

Figure 10.21 (a) Mean PIB data collected at 6:30-7:30 pm on June 7, 2011 at the Fremont end of the 17-km link and (b) mean PIF data during the same time. PIB data is recorded at 7.68 kHz with 45 s averages whereas PIF data is recorded at 20 kHz with 45 s averages. The slanted-line areas represent times when the AO was turned on at both ends of the link. The dotted area represent times with Tx AO on and Rx AO off, and the black box represents times with Rx AO on and Tx AO off.

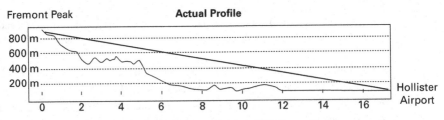

Figure 10.22 Ground profile between Fremont Peak and Hollister Airport.

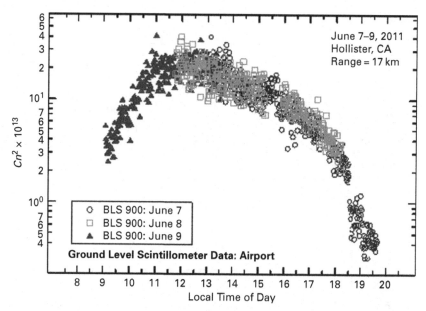

Figure 10.23 Ground level refractive index structure parameter (C_n^2) values at Hollister Airport during June 7-9, 2011 measured by a BLS 900 Scintec scintillometer.

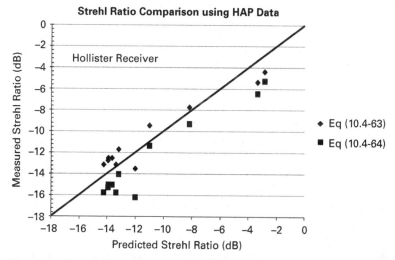

Figure 10.24 Comparison of measured and predicted Andrews/Phillips Strehl ratio over a 17 km link for June 8, 2011 using HAP model.

very encouraging results, especially when compared to the Roddier and Noll theory with JHU APL data. Figure 10.25 depicts the same comparison as in Figure 10.24, except we use the Roddier and Noll theories. Here, the Roddier Strehl ratio is given by the Roddier gain equation, Eq. (10.55), multiplied by the Andrews/Phillips Strehl ratio, Eq. (10.43). These plots clearly show that both theories expect the AO systems to mitigate essentially all effects under all turbulence conditions, which the data in the plot

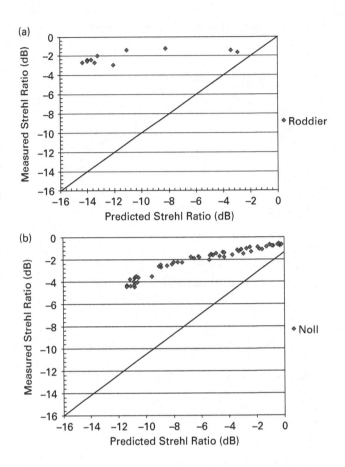

Figure 10.25 Comparison of measured and predicted (a) Roddier and (b) Noll Strehl ratios using the HAP model and 17 km link data taken on June 8, 2011.

do not support. The conclusion is that one should not use these theories outside the weak turbulence conditions.

The bottom line is that both Roddier's and Noll's equations significantly overestimate the ability of the AO system to mitigate turbulent effects. The Andrews/Phillips theory does a better job. All of the above make linear modification to the Strehl ratio, but the turbulent channel is really non-linear. Maybe, a more non-linear approach is needed.

10.6 The ORCA/FOENEX scintillation mitigation model

The Optical RF Communications Adjunct (ORCA) and Free-space Optical Experimental Network EXperiment (FOENEX) developed a new methodology of estimating the FSOC link budgets for a large range of optical turbulent conditions [34–36]. The approach comes from the observation that the ability of an AO system to compensate turbulence along a path is limited by the transmitter and receiver Rayleigh range, proportional to the diameter of the optics-squared and inverse of the wavelength of light utilized. The method uses the Fried parameter computed over the range outside the transmitter and receiver Rayleigh ranges, to calculate the Strehl ratios that yield a reasonable prediction of the light impinging on the receiving telescope's aperture and the power coupling into the fiber. This section develops this model and shows a comparison of its estimates with various sets of experimental data.

An optical channel model must first, and foremost, account for the spreading of the beam between the transmitter and receiver. In free space, the Fraunhofer equation provides the far-field irradiance at the receiver aperture as a function of range, R, wavelength, λ, and transmitter aperture area, A_{tx}, as

$$I_{aperture} \approx \gamma_{tx} P_{tx} \left(\frac{A_{tx}}{(\lambda R)^2} \right) \tag{10.65}$$

where γ_{tx} is the transmitter optics transmittance, P_{tx} the laser transmitter power, and $A_{tx} = \frac{\pi D_{tx}^2}{4}$ the area of the transmitter aperture [37].[2] The power at the aperture is then this irradiance multiplied by the area of the receiver

$$P_{rx} \approx \gamma_{tx} P_{tx} \left(\frac{A_{tx} A_{rx}}{(\lambda R)^2} \right) \tag{10.66}$$

$$\approx \gamma_{tx} P_{tx} FSL \tag{10.67}$$

where $A_{rx} = \frac{\pi D_{rx}^2}{4}$ is the area of the receiver aperture and

$$FSL \approx \frac{A_{tx} A_{rx}}{(\lambda R)^2} \tag{10.68}$$

is the Fraunhofer spreading loss (FSL) [34].

One aspect of the AO system not discussed explicitly is that the above assumes the AO sensor measures all the turbulent effects over the entire laser link. Given Nyquist sampling considerations using a finite aperture, this cannot happen. Stotts *et al.* hypothesized that the WFS could only measure the optical distortion from either the transmitter or receiver over a small distance called the Rayleigh range (RR) [35,36]. In other words, the AO system cannot mitigate turbulence effects beyond the Rayleigh ranges from either end. Let us look at this in more detail.

The Rayleigh range is defined as the distance along the propagation direction of a beam from the beam waist to the location where the area of the cross section is doubled. A laser communications system is assumed with a beam exiting a telescope of diameter D. From diffraction theory, we know that range R where the beam is twice the cross section of the exit diameter can be derived from the relationship:

$$\sqrt{2}D \approx \frac{\lambda R}{D} \tag{10.69}$$

---

[2] The formula was derived in 1945 by Danish-American radio engineer Harald T. Friis at Bell Labs [36].

where $D \equiv$ diameter of the telescope, $\lambda \equiv$ wavelength of light used in the systems, $R \equiv$ range.

If one wants to Nyquist sample the optical field at the range, R, then the imaging system needs to sample at a resolution equal to half this diameter. This implies that

$$\frac{D}{\sqrt{2}} \approx \frac{\lambda R}{D}$$

or

$$R \approx \frac{D^2}{\sqrt{2}\lambda} = \frac{0.7D^2}{\lambda} \tag{10.70}$$

where D^2/λ is known as the Fresnel distance or Fresnel length from diffraction theory. For $D = 0.1$ meters and eye-safe laser light, we have

$$R \sim 0.7 \,(0.1 \text{ m})^2 / (1.55 \times 10^{-6}\text{m})$$
$$= 4.5 \text{ km}. \tag{10.71}$$

Another way to look at it is that the image of AO effects at a range R is of extent D. The image of spherical, coma, and astigmatism is of extent $D/2$ or $D/3$. Therefore, the range at which to resolve (a lenslet) of that size is $D^2/2\lambda$ or $D^2/3\lambda$ This agrees with the sampling theorem reasoning given above that the maximum sample range is of the order of the Fresnel distance.

When turbulence is present, Stotts et al. have shown that the above irradiance is further decreased by the transmitter and receiver Strehl ratios, SR_{tx} and SR_{rx}, respectively [21,22]. Specifically, they showed the power at the receiver aperture can be written as

$$P_{rx} \approx \gamma_{tx} P_{tx} \left(\frac{A_{tx} A_{rx} \, SR_{tx}}{(\lambda R)^2} \right) \tag{10.72}$$

The received power at the detector includes the receiver optics and fiber irradiance terms (γ_{rx} and γ_{fiber}) as well as the receiver Strehl ratio. It is written as

$$P_{detector} \approx \gamma_{tx} \gamma_{rx} P_{tx} \left(\frac{A_{tx} A_{rx} \gamma_{fiber} \, SR_{tx}}{(\lambda R)^2} \right) SR_{rx} \tag{10.73}$$

$$\approx \gamma_{tx} \gamma_{rx} \gamma_{fiber} \, P_{tx} \, FSL \, SR_{tx} \, SR_{rx} \tag{10.74}$$

The $\left(\frac{D}{r_0} \right)$ term is dependent upon the reference to the receiver or transmitter, giving $\left(\frac{D_{rx}}{r_{0rx}} \right)$ and $\left(\frac{D_{tx}}{r_{0tx}} \right)$ respectively. The corresponding receiver coherence lengths are shown in [20,21] to be

$$r_{0rx} = \left[16.71 \int_{R_{txRR}}^{L-R_{rxRR}} C_n^2(r)(r/R)^{5/3} \frac{dr}{\lambda^2} \right]^{-3/5} \tag{10.75}$$

and

$$r_{0_{tx}} = \left[16.71 \int_{R_{txRR}}^{L-R_{rxRR}} C_n^2(r)(1-r/R)^{5/3} \frac{dr}{\lambda^2} \right]^{-3/5} \tag{10.76}$$

R_{rxRR} and R_{txRR} are the Rayleigh ranges for the receiver and transmitter, respectively [36].

For FSO transmissions that vary in altitude, the model must integrate across the path length to account for the various attenuation coefficients described in section 10.1. Here, the power P_{rx} is the PIB and the power $P_{detector}$ the PIF [18,21,22]. In their analyses, the PIB and PIF represent the median value of their associated statistical fading distributions [18,21,22]. The resulting power PIB, considering the factors discussed to this point, is given by

$$P_{rx} \approx \gamma_{tx} P_{tx} \left(\frac{A_{tx} A_{rx} \, \text{SR}_{tx}}{(\lambda R)^2} \right) \int_{h_{min}}^{h_{msx}} e^{-cR(h)} \, dh \tag{10.77}$$

and the resulting PIF is given by

$$P_{detector} \approx \gamma_{tx} \gamma_{rx} P_{tx} \left(\frac{A_{tx} A_{rx} \gamma_{fiber} \, \text{SR}_{tx}}{(\lambda R)^2} \right) \text{SR}_{rx} \int_{h_{min}}^{h_{msx}} e^{-cR(h)} \, dh \tag{10.78}$$

Figure 10.26 depicts the cumulative density function (CDF) as a function of PIF from references [30,35,36]. It is clear in this log-log plot that the median level and all higher order statistics are linearly related. The same relation holds for the PIB. This indicates that we should be able to derive a linear mapping between the 99% and median levels for these two measurements. Specifically, Stotts *et al.* developed the following linear relationships between the median PIF and the 99-percentile PIF [30,36]:

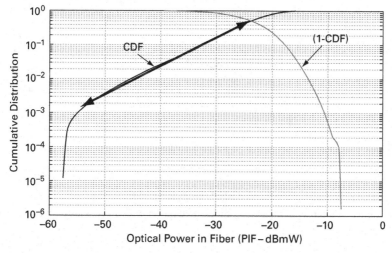

Figure 10.26 The statistical distribution of the PIF with adaptive optics produces a consistent linear relationship between log optical power and log cumulative distribution.

$$\text{PIF}(99\%) \approx 0.96 \, \text{PIF(median)} - 18.8 \, \text{dBmW} \pm 1.71 \, \text{dBmW} \qquad (10.79)$$

for urban/rural environments and

$$\text{PIF}(99\%) \approx 0.96 \, \text{PIF(median)} - 21.5 \, \text{dBmW} \pm 1.71 \, \text{dBmW} \qquad (10.80)$$

from the maritime environments [30].[3]

Given the model described above, a new methodology for estimating the FSOC link budget for atmospheric conditions ranging from weak to severe has been developed based on field trial data and is described in references [30,35]. Table 10.4 is an example link budget spreadsheet for the ORCA NTTR and PAX field trials, with measured data [30,35,36]. Figure 10.27 is a graphical comparison between the above theory and field measurements from these and two other sites [30,35,36]. For this plot, the model utilized a Rayleigh range of 4.5 km and multiples of the HV 5/7 model based on the refractive index structure parameter data plots in Figures 10.28(a) and 10.29(b). Figure 10.29 plots the measured and observed scintillation for the cited tests in Figure 10.29 onto Figure 10.9. This figure clearly shows that all of the reported results are under saturation regime conditions up to extreme levels.

Figure 10.30 is a comparison between the above theory and field measurements for a 17.4 km link between Hollister and Fremont Peak, both in Northern California [35,36]. These calculations used a Rayleigh range of 4.5 km and multiples of the HV 5/7 based on the C_n^2 data plots in Figure 10.23 [36]. Again, good agreement is depicted.

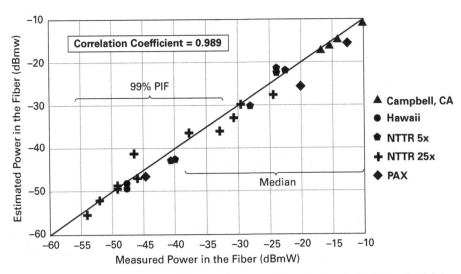

Figure 10.27 Summary comparison of measured power with predictions derived from Rayleigh Range Link Budget model.

---

[3] In references [30], the authors used the loss value of 19.5 dB as an engineering approximation of the above equation. However, the difference between the equation value and 19.5 dB is small, and the results are very similar.

Table 10.4 Link budgets for ORCA NTTR and PAX field trials

Parameters	Units	NTTR 5× HV5/7 A2AL 1:54:06	NTTR 5× HV5/7 A2AL 1:54:19	NTTR 25× HV5/7 AH2AL Value	NTTR 25× HV5/7 AH2AL Value	NTTR 25× HV5/7 AH2AL Value	NTTR 25× HV5/7 AH2AL Value	NTTR 25× HV5/7 AH2AL Value	NTTR 25× HV5/7 AH2AL Value	PAX 12 May 1× HG5/7 AH2AL Value	PAX 12 May 1× HG5/7 AH2AL Value
Tx aperture diameter	m	0.1	0.1	0.1	0.1	0.1	0.1	0.1	0.1	0.1	0.1
Tx altitude	1000 ft	26	26	26	26	26	26	26	26	26	26
Range	km	170	170	100	140	138	150	170	192	50	70
Rx aperture diameter	m	0.1	0.1	0.1	0.1	0.1	0.1	0.1	0.1	0.1	0.1
Rx altitude	1000 ft	7.8	7.8	7.8	7.8	7.8	7.8	7.8	7.8	0	0
Calculated Tx r_0 reduced by RR	cm	13.64	13.64	7.65	5.99	6.04	5.72	5.19	4.75	28.26	21.64
Calculated Rx r_0 reduced by RR	cm	8.46	8.46	5.10	3.81	3.86	3.62	3.22	2.90	14.48	8.24
Saturated Tx Power	dBmW	39.8	39.8	37.8	37.8	39.8	39.8	39.8	39.8	35.2	35.2
Tx feed loss (fiber/circulator)	dB	-3.6	-3.6	-3.6	-3.6	-3.6	-3.6	-3.6	-3.6	-3.6	-3.6
Tx feed loss (optics)	dB	-3.4	-3.4	-3.4	-3.4	-3.4	-3.4	-3.4	-3.4	-3.4	-3.4
Power at the Tx aperture	dBmW	32.7	32.7	30.7	30.7	32.8	32.8	32.8	32.8	28.2	28.2
Tx TBL loss	dB	-4.0	-4.0	-4.0	-4.0	-4.0	-4.0	-4.0	-4.0	-4.0	-4.0
Fraunhofer spreading loss	dB	-30.5	-30.5	-25.09	-28.8	-28.7	-29.4	-30.5	-31.6	-19.9	-22.8
TX Strehl loss	dB	-2.6	-2.6	-4.9	-6.3	-6.2	-6.2	-7.2	-7.8	-0.8	-1.3
Atmo absorption loss	dB	-4.7	-4.7	-2.7	-3.8	-3.7	-4.1	-4.7	-5.4	-3.5	-4.9
Rx TBL loss	dB	0.0	0.0	0.0	0.0	0.0	0.0	0.0	0.0	0.0	0.0
Filter loss	dB	0.0	0.0	0.0	0.0	0.0	0.0	0.0	0.0	0.0	0.0
Medium power in bucket	dBmW	-9.0	-9.0	-6.7	-12.2	-9.9	-11.3	-13.7	-16.0	-0.1	-4.8
Rx TBL loss	dB	0.0	0.0	0.0	0.0	0.0	0.0	0.0	0.0	0.0	0.0
Rx feed loss (optics)	dB	-7.8	-7.8	-7.8	-7.8	-7.8	-7.8	-7.8	-7.8	-7.8	-7.8
Rx feed loss (fiber/circulator)	dB	-2.5	-2.5	-2.5	-2.5	-2.5	-2.5	-2.5	-2.5	-2.5	-2.5
Rx Strehl loss	dB	-4.8	-4.8	-7.3	-9.3	-9.2	-9.7	-10.6	-11.4	-2.2	-4.5
Medium power in the fiber	dBmW	-24.1	-24.1	-24.3	-31.7	-29.4	-31.3	-34.5	-37.6	-12.5	-19.6
Measured PIF	dBmW	-22.5	-24.0	-24.0	-28.0	-24.5	-29.5	-30.5	-38.0	-13.0	-20.0
PIF fade to 99% value	dB	19.0	19.0	19.0	19.0	19.0	19.0	19.0	19.0	21.5	21.5
Power in the fiber at 99%	dBmW	-42.1	-42.1	-42.3	-49.5	-47.2	-49.0	-52.1	-55.1	-33.5	-40.3
Measured PIF at 99%	dBmW	-40.5	-40.0	-46.5	-49.0	-46.0	-49.0	-52.0	-54.0	-33.0	-45.0

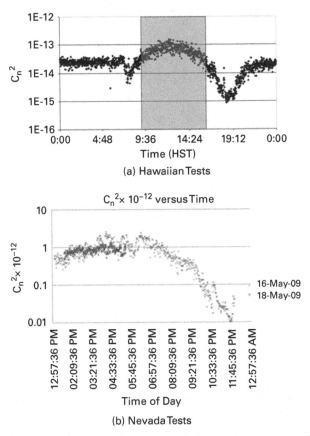

Figure 10.28 Summary of the refractive index structure parameter (C_n^2) measurements as function of the time of day for three experimental sites.

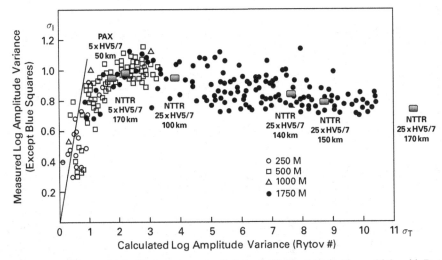

Figure 10.29 Observed scintillation versus predicted scintillation Plot in Figure 10.9 with Rytov numbers and measured log variance from Hawaii, Nevada and California Tests.

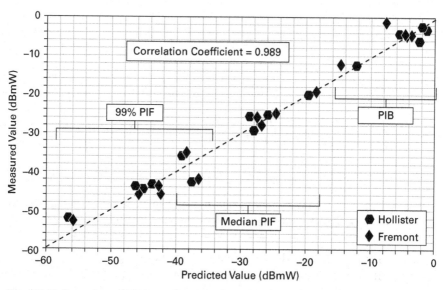

Figure 10.30 Summary comparison of measured power with predictions derived from Rayleigh Range Link Budget model.

In 1969, Fried showed the equivalence of transmitter and receiver gain ("reciprocity") in free space and in random media [25], which was expanded by Kennedy and Karp [37], Shapiro [38] and others [39,40]. This is validated by the results in Figure 10.30 for the Hollister and Freemont data. In the last two references, Walther [39] and Parenti *et al.* [40] showed a strong correction was between the air and ground PIFs.

Example 10.3 During 11–13 May 2011, the FOENEX program took PIB and PIF data for a 10 km link running from Saratoga, California, to Campbell, California. Figure 10.22 again shows the refractive index structure parameter for the 3-day experiment. In example 10.1, we calculated the link slant path angle for their link geometry, which turned out to be 87.14 degrees. For simplicity, we will assume this is essentially a horizontal path.

The receiver and transmitter Fried parameters for a horizontal link are given by

$$r_0^R = \left(16.71\, C_n{}^2(0)\, \lambda^{-2} \int_0^R \left(\frac{r}{R} \right)^{5/3} dr \right)^{-3/5}$$

$$= \left(16.71\, C_n{}^2(0)\, \lambda^{-2} R \int_0^1 (z)^{5/3} dz \right)^{-3/5}$$

$$= \left(16.71\, C_n{}^2(0)\, \lambda^{-2} R \left. \frac{z^{8/3}}{8/3} \right|_0^1 \right)^{-3/5}$$

$$= \left(6.266\, C_n{}^2(0)\, \lambda^{-2} R \right)^{-3/5}$$

and

$$
\begin{aligned}
r_0^T &= \left(16.71\, C_n{}^2(0)\, \lambda^{-2} \int_0^R \left(1 - \frac{r}{R}\right)^{5/3} dr \right)^{-3/5} \\
&= \left(16.71\, C_n{}^2(0)\, \lambda^{-2} R \int_1^0 (1-z)^{5/3}\, dz \right)^{-3/5} \\
&= \left(16.71\, C_n{}^2(0)\, \lambda^{-2} R\, \frac{(1-z)^{8/3}}{8/3} \bigg|_1^0 \right)^{-3/5} \\
&= \left(6.266\, C_n{}^2(0)\, \lambda^{-2} R \right)^{-3/5}
\end{aligned}
$$

respectively. Normally, one would use the entire range for this calculation. However, ORCA/FOENEX model integrates from the RR to R-RR. Let us show a sample calculation.

Let us look at the link test held at 0645. Referring to Figure 10.22, we estimate the refractive index structure parameter to equal to $C_n{}^2(0) \approx 3\, \tilde{n}\, 10^{-14}\, m^{-2/3}$. For a wavelength of 1.55 μm, transmitter/receiver aperture diameters of 10 cm and this constant refractive index structure parameter, we would calculate a transmitter/receiver Fried parameter equal to 7.3 cm using the above equations and a integration range of 1 km (10 km – 2 × RR = 10 km – 2 × 4.5 km = 1 km). The resulting Strehl ratio is 0.306 or −5.1 dB. Table 10.5 illustrates the link budget for the May tests at that time.

Beside average PIB and PIB data for that experiment, Juan Juarez, JHU APL, provide median values for selected times and dates from this experiment [33]. Table 10.6 shows the link budget median PIB and PIF estimates with measured data, which is graphed in Figure 10.31. This shows fairly good agreement for the constant horizontal link assumption.

10.7   Diversity scintillation mitigation techniques

The Free-Space Optical Communications Airborne Link (FOCAL) programs and associated research effort conducted by researchers from MIT Lincoln Laboratory have used spatial diversity, forward error correction (FEC) and interweaving to reduce fading loss, projected to be of the order of 20 dB [39,42–54]. This section will summarize their basic approach and results [39].

Optical diversity ideally reduces the log fade level by the number of apertures if beams are statistically independent; this experiment employed four receiver apertures. Achieving that independence requires separating the beams by a distance greater than the Fried parameter [39,53]. This is the first scintillation technique used by the FOCAL program. The researchers used an aircraft aperture of 2.54 cm with four ground 2.54 cm apertures, varying the separation from each other by 10–48 cm. The downlink signal is separately detected in each aperture chain, with signals then summed for clock recovery and bit detection. Receiver filter bandwidth is 10 GHz. A dynamic variable optical attenuator (VOA) is included in each detection circuit to stabilize signal levels at the decision circuits. A system block diagram and examples of collected data are shown in Figure 10.32.

Table 10.5 Link budgets for May 2011 10-km field trials

Parameters	Units	Samtoga GND 1 Rcvr no turbulence	Campbell AIR 1 Rcvr no turbulence
Tx aperture diameter	m	0.1	0.1
Tx altitude	1000 ft	0.1	0.1
Range	km	10	10
Rx aperture diameter	m	0.1	0.1
Rx altitude	1000 ft	0.94	0.94
Calculated Tx r_0 reduced by RR	cm	1.94	2.91
Calculated Rx r_0 reduced by RR	cm	1.91	2.94
Saturated Tx power	dBmW	23.0	23.0
Tx feed loss (fiber junction box)	dB	−1.6	−1.5
Tx feed loss (optics)	dB	−4.9	−4.3
Power at the Tx aperture	dBmW	16.5	17.2
Tx TBL loss	dB	0.0	0.0
Franhofer area at Rx aperture	dBsm	−15.1	−15.1
Tx aperture area	dBsm	−21.0	−21.0
Fraunhofer spreading loss	dB	−5.9	−5.9
TX Strehl loss	dB	−5.1	−5.1
Atmo absorption loss	dB	−1.3	−1.3
Rx TBL loss	dB	0.0	0.0
Jitter loss	dB	0.0	0.0
Median power in bucket	dBmW	4.1	4.8
Rx TBL loss	dB	0.0	0.0
Rx feed loss (optics)	dB	−6.1	−5.2
Rx feed loss (fiber junction box)	dB	−1.9	−1.6
Rx Strehl loss	dB	−5.1	−5.1
Median power in the fiber	dBmW	−9.0	−7.1

The second scintillation mitigation technique used by the team was FEC coding with interleaving; the basic interleaving concept is illustrated in Figure 10.33. The FEC adds symbols to each code word, allowing recovery of all symbols if some are lost to fading. The Reed-Solomon (255,239) code used can correct 8 byte (symbol) errors per code word. For OTU1 framing with RS (255,239) FEC, 64 code words of 255 bytes each constitute a ~50 µs frame, far shorter than a millisecond class atmospheric fade. If all symbols of the code word fall in the fade, the FEC is ineffective. Interleaving provides temporal diversity by spacing the code symbols by a time duration in excess of the characteristic atmospheric fade. For this work, the symbols were spaced by 5 ms, leading to a latency of 1.25 s after de-interleaving. The combination of an optical mitigation scheme to reduce scintillation fading and an FEC code operating with sufficient interleaving in the data domain can make communications in the deep fading channel tractable and significantly extend the available range for air-to-ground communications. Figure 10.34 illustrates the improvement achieved in the mean power level (referred to as the static loss) and the reduction of the signal fluctuations (referred to as the dynamic loss). Under severe turbulence conditions, the combined effect of spatial diversity, interleaving, and encoding can easily reduce the transmitted signal requirement by more than 20 dB.

Table 10.6 Link budgets for May 2011 Saratoga-Campbell field trials

Date	Local time	$C_n^2(0)$ estimate	Fried parameter	Saratoga, CA				Campbell			
				Estimate PIB	Measure PIB	Estimate PIF	Measured PIF	Estimate PIB	Measured PIB	Estimate PIF	Measured PIF
12 May 2011	18:45	3.0×10^{-14} m$^{-2/3}$	7.3 cm	4.1	4.71	−9	−4.84	4.8	5.47	−7.1	−3.2
13 May 2011	06:45	3.0×10^{-14} m$^{-2/3}$	7.3 cm	4.1	4.69	−9	−5.9	4.8	2.81	−7.1	−4.6
13 May 2011	10:00	3.0×10^{-13} m$^{-2/3}$	2.3 cm	−3.8	−3.38	−24.8	−22.89	−3.1	−5.43	−22.4	−21.77
13 May 2011	13:45	3.0×10^{-13} m$^{-2/3}$	1.8 cm	−5.7	−8.44	−28.7	−26.7	−5	−4.43	−26.8	−25.92

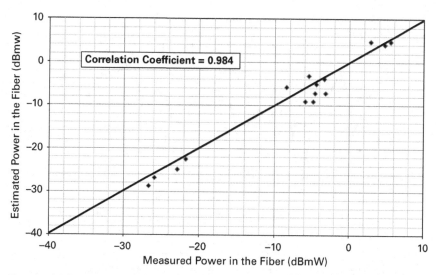

Figure 10.31 Summary comparison of measured power with predictions derived from a constant Rayleigh Range Link Budget model.

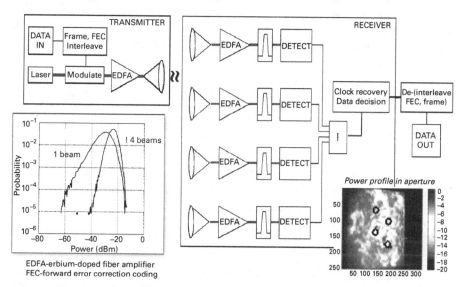

Figure 10.32 The downlink path for the FOCAL experiment employed a single transmitter mounted in the aircraft and four independently-tracked detectors in the ground-based receiver. The receiver outputs were incoherently summed, and the resulting improvement is demonstrated by the distributions shown in chart on the lower left-hand side of this figure. The detector layout and an image of the scintillation pattern is shown in the picture on the lower right-hand side of this figure [51].

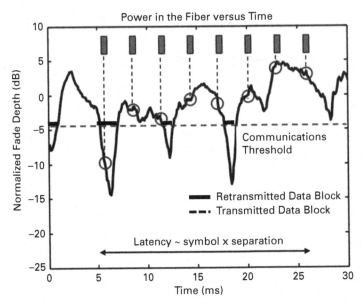

Figure 10.33 Interleaving refers to a technique in which a code block is segmented into a string of symbols that are temporally separated in order to minimize the likelihood that an entire block will be transmitted during a channel fade. The use of encoding allows the entire code block to be recovered even if some of the symbols are decoded incorrectly [51].

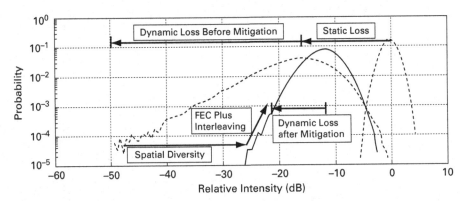

Figure 10.34 Comparison of received signal distribution functions for a short link with low turbulence (dashed, RH curve), a unmitigated long link with high turbulence (dashed, LH curve, and the long link with scintillation mitigation (solid curve). Optimized mitigation schemes can reduce the transmitted signal power requirement by more than 20 dB. [51]

10.8   An optical modem model

In the previous two sections, we have described several approaches for scintillation mitigation, some of which have shown around 20 dB of link improvement [34–36]. This section will talk about the use of an optical modem that provides the significant receiver

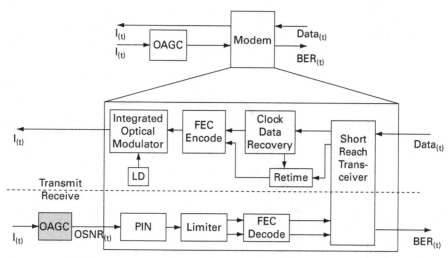

Figure 10.35 Optical modem block diagram.

sensitivity / systems gain for mitigating turbulent effects in excess of 20+ dB. It has helped the ORCA/FOENEX program provide error-free and near error-free communications for FSOC links operating in saturation regime turbulences where additional of 20–30 dB occurred after the AO systems, e.g. > 45–55 dB unmitigated channel fading. Figure 10.35 shows the basic architecture for the optical modem. It is composed of two stages. The first stage is called the automatic optical gain control (OAGC) and the second is a FEC capability [55–57]. Juarez *et al.* have configured these two stages to create an optical modem designed to convert the widely-varying optical signal into a stream of digital data for processing by a network router [55–57]. They can provide 25–30 dB of link performance improvement.

The OAGC must perform several functions to stabilize the highly fluctuating received signal for reliable detection. The first is protection of the photodiodes and follow-on electronics from saturation or catastrophic damage due to high optical power levels. This condition can exist in FSOC links over short distances (< 10 km) or even on longer (> 100 km) links during benign turbulence. Additionally, architectures employing fixed gain optical pre-amplifiers, such as erbium-doped fiber amplifiers (EDFAs), are especially susceptible because they can output power levels well above the damage thresholds of detectors in response to rapid power transients in a "Q-switch" effect. The second function the OAGC must perform is to provide low-noise optical amplification to improve receiver sensitivity. This is critical to maximize the communications link margin. Lastly, the OAGC must reduce the power transients that couple through the receiver follow-on electronics and degrade bit error rate performance. The OAGC achieves this by optically amplifying or attenuating as necessary through a series of multiple stages as discussed in [29,30] to output a constant power (POF) at a level of optimal performance for the detector. In essence, the time-variant optical input [$I(t)$] is translated into a constant output with a variable optical signal-to-noise ratio [OSNR(t)]. This performance is illustrated in Figure 10.36, which presents OSNR and OAGC output power as a function of power into the OAGC (which is

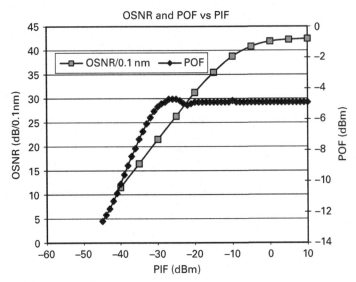

Figure 10.36 Optical signal-to-noise ratio (OSNR) and power out of the fiber (POF) versus input power in the fiber (PIF).

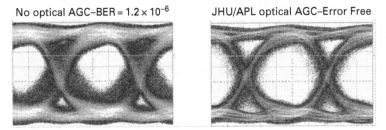

Figure 10.37 Eye diagrams for FSOC link with, and without, the OAGC.

equivalent to PIF). The signal modulation scheme was NRZ-OOK. The maximum gain of the first-generation system is shown to be between 40 and 45 dB. Figure 10.37 shows the eye-diagram for the FSOC link with and without the OAGC. This figure clearly illustrates an improvement in signal quality with the use of the OAGC, which resulted in reducing the BER from 2.1×10^{-6} to $< 10^{-12}$ in that particular experiment [26]. In discussions to come, POF stands for power out of the fiber (OAGC).

The modem was designed to interface between the 10 Gbps Ethernet client and the 11 Gbps FSO line rate. Specifically, the modem uses a commercial off-the-shelf (COTS) Reed-Solomon (255,239) enhanced FEC, with a 7% overhead chipset for optical links, which are designed to operate in a high received power, variable OSNR environment. Lab tests have proven that COTS FEC chips can provide the full designed 8 dB of gain, even when the power into the OAGC varies over 4 orders of magnitude. The primary penalty to outages below system sensitivity is the time the FSO side clock recovery circuitry takes to acquire clock after a fade. This was characterized in the lab to be of the order of 100 μs. The BER

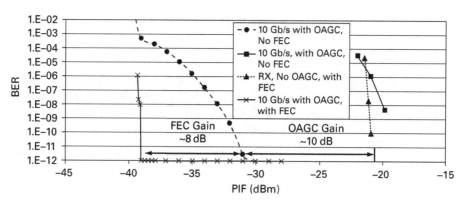

Figure 10.38 Bit error rate versus input power in the fiber (PIF) with OAGC only, and with OAGC and FEC.

results of using the COTS FEC with the first-generation OAGC are shown in Figure 10.38. For this test, the BER performance of the back-to-back test was compared with and without the OAGC using InGaAs PIN/TIA receivers for both cases. For the case without the OAGC, low received optical power starves the photo-receiver and leads to decision errors by the limiting amplifier, which cannot be corrected by the FEC. For this case, the FEC provides little gain as the low receiver power quickly reduces the received signal-to-noise ratio. This is not surprising as the FEC chipset was designed for optical fiber communications systems with optical pre-amplification where link configurations and OSNR and power levels are carefully controlled.

Figure 10.39 illustrates how the optical modem performed on a 183 km link during late afternoon conditions (4–6 p.m. in Figure 10.28(b)]. In particular, this figure shows PIF data taken on 18 May 2009 during Flight 2, with POF data taken during laboratory testing. The turbulent conditions were estimated at 5× HV 5/7. The optical modem developed for these tests utilized an NRZ-OOK modulation format, operating at a line rate of 10.3125 Gbps. The FEC used was low overhead (7%) Reed-Solomon code cited above. Combined with the OAGC originally demonstrated in 2007, the PIF noise floor (10^{-12} BER) was expected to be −39 dBmW [26], but from Figure 10.39 we see that that OAGC was able to maintain a constant output power over a greater PIF range, thus lowering the 10^{-12} BER point to −41 dBmW. This was due to a minor system upgrade prior to Nevada field test.

Juarez *et al.* have reported the improved operation in the upgraded OAGC discussed above [56]. A sensitivity of −410.9 dBmW, i.e., 10 photons per bit, at 10.7 Gbps was achieved employing a return-to-zero DPSK-based modem, which includes improved OAGC and the Reed-Solomon enhanced FEC (255,239) (7% OH). Low-noise optical gain for the system was provided by an OAGC with a noise figure of 4.1 dB (including system required input loses) and a dynamic range of greater than 60 dB. Figure 10.40 illustrates the result shown in Figure 10.38 with the improved BER versus PIF with the new signal modulation scheme (~ 3 dB improvement) and other modifications. This figure shows an additional ~ 8 dB improvement in performance from these changes.

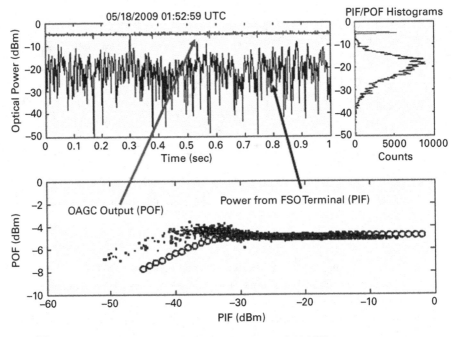

Figure 10.39 Performance of OAGC on 18 May NTTR data, with comparison with laboratory characterization results.

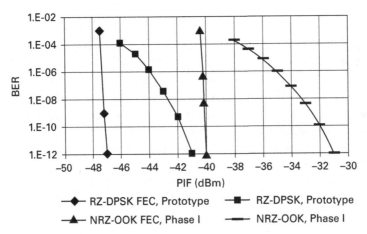

Figure 10.40 Updated Figure 10.38 comparing original non-return-to-zero on-off key (NRZ-OOK) signal modulation with new return-to-zero differential phase shift key (NZ-DPSK) signal modulation.

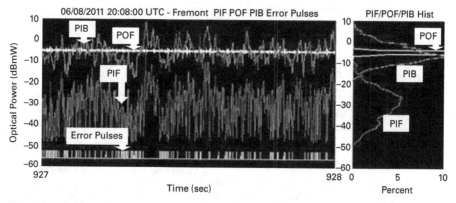

Figure 10.41 Performance of optical modem on June 8, 2011 [57].

Figure 10.41 illustrates how the updated optical modem performed on the 17.4 km Hollister-Fremont Peak (California) link at 1 p.m. local time, 8 June 2011 (Figure 10.23). In this figure, the error pulses are < 2% packet error rate. This is really good performance as the turbulent conditions were extreme, twice what was experienced in Nevada [35,36].

So how would one create error-free communications in this and worse conditions as we have reached the limit of the optical modem? Stotts *et al.* proposed a hybrid optical-RF network for this purpose [26,36,57]. In the situation illustrated by Figure 10.41, the 300 Mbps RF systems could transmit the lost packets depicted by the error pulses as 2% of 10 Gbps is 200 Mbps [26,36]. However, for larger losses, the network would need to handle these errors. For example, Stotts *et al.* showed a network router could handle 5% and 15% packet losses using a modified (for increased loss) data stream taken under the IRON-T2 program [57].

Figure 10.42 shows the high-level system architecture for the network. Here, IPCM indicates the Inter-Platform Communications Manager, which is the FOENEX MANET, topology manager, and network controller. XIA refers to the XFUSION Interface Assembly (XIA is the third generation (first generation was IRON-T2, second was ORCLE/ORCA Phase 1 XFUSION) hybrid-aware network router and controller developed for ORCA/ FOENEX) [57]. In this configuration, the networking subsystem is realized in the XIA hybrid-aware IP router and provides seamless integration of the RF and FSO subsystems to form hybrid communication links utilizing the attributes and strengths of each channel to meet the network availability and QoS requirements for the FOENEX network. Note that the IPCM is the FOENEX Mobile Ad hoc NETwork (MANET) and network control software that resides on the FOENEX XFUSION Interface Assembly (XIA) hybrid-aware router.

Figure 10.43 depicts the FOENEX Network Stack. The FOENEX network system and associated network stack utilizes a layered approach to improving network availability. At the physical layer the system utilizes the concept of hybrid links (the bonding of RF and FSO channels) for use in communications between one network node and another. These hybrid links provide techniques for link layer retransmission, prioritized failover of traffic flows from one hybrid channel to another and the ability to differentiate between atmospheric scintillation and cloud events from hybrid link outage. These link layer capabilities are tightly integrated with Diffserv QoS capabilities, local deep queuing

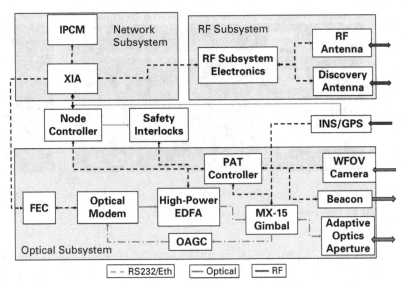

Figure 10.42 FOENEX high-level system architecture [57].

Capabilities of the FOENEX Network stack

– Network discovery, formation/reformation through real-time network control (discovery subsystem and Inter-Platform Communications Manager (IPCM))
– Hybrid link management and control
– Link outage mitigation through predictive link outage, Layer 2 re-transmission for handling scintillation effects, deeper queues for 2–3 second cloud blockages and replay of data for 5 second outages
– Mobility management to dampen the effects of mobility on standard Internet protocols (IPCM Adaptation Layer (IAL))
– Integrated Diffserv QoS for priority and internal and external ORCA network users, traffic management and flow control

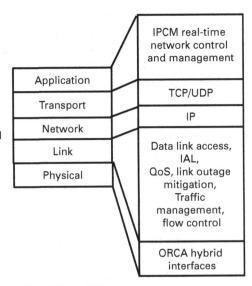

Figure 10.43 ORCA/FOENEX Network stack capabilities and breakdown [57].

techniques, and the ability to segregate network traffic flows into internal (primary mission users) and external (secondary mission users) traffic aggregates. The FOENEX MANET provides network node discovery for network formation/re-formation and reactive and proactive management of the network topology based on anticipated or incurred network outage events. Stability in the network is provided by dampening the effects of mobility on routing and higher-layer protocols and applications through the use of Layer 2 switching techniques (e.g., IAL).

Table 10.7 Test conditions for the network performance testing

Conditions of test
- Protocol = IPv4/1Pv
- Packet sizes = 1518 bytes
- Configuration = hybrid (5 Gbps user information on FSO, 200 Mpbs RF)
- Duration = 30 minutes per test
- Scintillation profile = 5% Hawaii (20 min repeating segment) using hardware scintillator
- L2 Retransmission – disabled vs. enabled
- L2 Retransmission timeout = 1 second

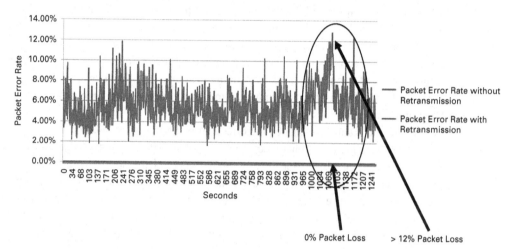

Figure 10.44 Network packet loss results for 5% level atmospheric turbulence, with and without Layer 2 retransmission [57].

To see if the network could handle 2% or higher PER traffic, network performance was assessed for the above FOENEX hardware and software using scintillation data from the AFRL IRON-T2 Hawaii experiments in October 2008 [26,36]. The experimental configuration consisted of two network nodes simultaneously transferring data each way, each containing a hybrid link connected to the another. This configuration will test network retransmission effectiveness. The FSO channel was at 10 Gbps clock rate carrying 5 Gbps of user data sourced from the IXIA. The RF channel is at a clock rate of 274 Mbps with a user information rate of 200 Mbps. Table 10.7 shows the test conditions for the network performance testing. Figure 10.44 shows the network performance for a 5% scintillation profile, with and without Layer 2 retransmission. It is clear from this figure that the saturation-regime atmospheric turbulence occurring during the IRON-T2 Hawaii experiments created significant packet loss, >12% at times. The RF link was used for retransmission requests. This figure shows error-free packet delivery using a hybrid link-based Layer 2 retransmission packet loss mitigation technique. In other words, there was not any degradation in data integrity errors, or dropped packets, during the test sequence. The authors also discussed additional laboratory testing, which included 12+ hour stress testing and using a mix of packet size profiles. They cited similar results for those tests using both 5% and 15% scintillation profiles.

10.9   Coherent optical communications

From Chapter 6, we know that coherent systems bring a number of advantages theoretically to FSOC. They have been proposed for use in the atmospheric channels as well [58–64]. Fernandez and Vilnrotter have demonstrated coherent optical arrayed receivers for pulse-position modulation (PPM) signals in FSOC [60]. In the 1980s, homodyne PSK communications demonstrated shot-noise-limited operation with receiver sensitivity equal to 9 photo-electrons per bit ($\eta = 1$). This is much better than the pre-amplified direct detection systems in the 1990s where signal-ASE beat-noise-limited operations were achieved with receiver sensitivities of 39 photo-electrons per bit ($\eta = 1$) for intensity-modulated direct detection and 20 photo-electrons per bit ($\eta = 1$) for DPSK. However, the system complexity of a coherent system made the latter approaches more desirable in real systems. Coherent systems started coming back in use by the mid-2000s, led by the work of Lange *et al.* for space communications [63]. The key to leveraging the coherent advantage is finding techniques to mitigate photo-mixing degradation because of amplitude and phase distortion from turbulence.

Belmonte and Kahn investigated the effects of wave-front distortion on the performance of synchronous (coherent) receivers utilizing phase compensation and diversity combining techniques [65]. Specifically, they compared the spectral efficiencies and power efficiencies of several modulation formats using coherent detection in the presence of multiplicative noise (fading) from atmospheric turbulence and additive white Gaussian noise (AWGN), assuming adaptive optics and diversity mitigation of the turbulent effects. In respect to the latter point, they assumed their receiver is able to track any temporal phase fluctuations caused by turbulence (as well as those caused by laser phase noise), performing ideal coherent (synchronous) demodulation. The result was receiver performance only required the knowledge of the envelope statistics of the down-converted electrical signal from the homodyne/heterodyne detection process.

Belmonte and Kahn defined the SNR per symbol per unit bandwidth as

$$\gamma_0 = \left(\frac{P}{N_0 B} \right) \tag{10.81}$$

assuming an AWGN channel. Again, P is the average power of the source, $N_0/2$ is the unilateral spectral noise density and B is the system bandwidth; γ_0 can be interpreted as the detected number of photo-electrons ($\eta = 1$), or photo-counts, per symbol when $1/B$ is the symbol period, just like in Chapter 6. In their analysis of a fading AWGN channel, they define α^2 to be the atmospheric channel power fading and the instantaneous received SNR per bit given in Eq. (10.81) now can be written as

$$\gamma_0 \alpha^2 = \left(\frac{P}{N_0 B} \right) \alpha^2 \tag{10.82}$$

In other words, the above SNR can be taken as the number of signal photo-electrons detected on the receiver aperture, multiplied by a heterodyne mixing efficiency α^2. When a coherent optical signal traverses the turbulent channel, the resulting signal will experience phase and amplitude fluctuations, or distortions. The spatial fields of the received signal will not properly match that of the local oscillator; hence, the contributions to the current signal from different parts of the receiver aperture can interfere destructively. This causes the reduced instantaneous mixing efficiency. The amount of this reduction is a function of D/r_0 and was shown in Figure 10.11.

Belmonte and Kahn stated that the statistical properties of the atmospheric random channel fading α^2 are related to the characteristics of atmospheric amplitude and phase fluctuations [65]. They wrote that the atmospheric log-amplitude fluctuations (scintillation) and phase variations (aberrations), can be characterized by their respective statistical variances,

$$\sigma_\chi{}^2 = \ln\left(1 + \sigma_\beta{}^2\right) \tag{10.83}$$

and

$$\sigma_\phi{}^2 = C_j \left(\frac{D}{r_0}\right)^{5/3}, \tag{10.84}$$

which define the impact of turbulence on mixing efficiency and fading. In Eq. (10.83), the intensity variance $\sigma_\beta{}^2$ represents the scintillation index. These equations imply that the statistics of phase aberrations caused by atmospheric turbulence are assumed to be characterized by a Kolmogorov spectrum of turbulence [65]. Noll showed that the classical results for the phase variance $\sigma_f{}^2$ could be extended to include modal compensation of atmospheric phase distortion [31]. In such modal compensation, Zernike polynomials are used similarly to the incoherent detection case as the basis functions because of their simple analytical expressions and their correspondence to classic aberrations [8]. The coefficient C_j depends on J [65]. A coefficient 1.0299 in the phase variance $\sigma_f{}^2$ assumes that no terms are corrected by a receiver employing active modal compensation. For example, aberrations up to tilt, astigmatism, coma and fifth-order correspond to $J = 3, 6, 10$ and 20, respectively. Ideally, it is desirable to choose J large enough that the residual variance $\sigma_f{}^2$ becomes negligible.

In addition, Belmonte and Kahn found that the SNR γ at the output of a perfect L-element diversity coherent combiner in the atmosphere would be described by a non-central chi-square distribution with $2L$ degrees of freedom:

$$p_\gamma\left(\gamma_{MRC}\right) = \left(\frac{1+r}{\bar{\gamma}}\right)^{\frac{L+1}{2}} \left(\frac{1}{Lr}\right)^{\frac{L-1}{2}} \exp(-Lr)$$

$$\times \exp\left(-\frac{(1+r)\,\gamma_{MRC}}{\bar{\gamma}}\right) I_{L-1}\left(2\sqrt{\frac{L\,(1+r)\,r\,\gamma_{MRC}}{\bar{\gamma}}}\right) \tag{10.85}$$

where $I_{L-1}(\ldots)$ is $(L{-}1)$-order modified Bessel function of the first kind. The average SNR (or average detected photo-counts) per symbol $\bar{\gamma}$ and the parameter $1/r$ describe turbulence effects [65, 66]. Both $\bar{\gamma}$ and $1/r$ are described in terms of the amplitude and phase variances $\sigma_\chi{}^2$ and $\sigma_f{}^2$ in Eqs. (10.83) and (10.84) [66]. The model leading to the PDF in Eq. (10.85) is based on the observation that the down-converted signal current can be characterized as the sum of many contributions from N different coherent regions within one aperture. In their model, the signal is characterized as the sum of a constant (coherent) term and a random (incoherent) residual halo. The contrast parameter $1/r$ is a measure of the strength of the residual halo relative to the coherent component. The parameter r ranges between 0 and ∞. It can be shown that when the constant term is very weak ($r \to 0$), turbulence fading causes the SNR to become gamma-distributed, just as in a speckle pattern [68]. Likewise, when the dominant term is very strong ($r \to \infty$), the density function becomes highly peaked around the mean value $\bar{\gamma}$, and there is no fading to be considered.

Belmonte and Kahn also stated that the model behind Eq. (10.85) can be applied to diversity schemes based on maximal ratio combining (MRC) of received signals. They pointed out that MRC schemes assume perfect knowledge of the branch amplitudes and phases, require independent processing of each branch and require that the individual signals from each branch be weighted by their signal amplitude-to-noise variance ratios, then summed coherently [67].

Assuming L independent branch signals and equal average branch SNR per symbol $\bar{\gamma}$, they cited that a receiver using MRC will weigh the L diversity branches by the complex conjugates of their respective fading gains and sum them. By setting $L = 1$, they pointed out that the PDF in Eq. (10.85) describes the SNR γ for a single receiving branch and corresponds to a non-central chi-square probability with two degrees of freedom [69].

Given the above, Belmonte and Kahn investigated several M-ary modulation techniques with coherent detection, including phase-shift keying (PSK), quadrature-amplitude modulation (QAM), and pulse-amplitude modulation (PAM). The parameter M is the number of points in the signal constellation and, consequently, $\log_2 M$ describes the number of coded bits per symbol. They also consider binary differential phase-shift keying (2-DPSK or DBPSK) with differentially coherent detection. Figure 10.45 plots the error probability of binary PSK in AWGN and in atmospheric fading from their analysis. Here is what they found. For $M = 2$, PAM and PSK are identical modulation formats. As should be expected, in AWGN the BER decreases exponentially with increasing photo-electrons per bit. However, under atmospheric fading, when no compensation techniques are considered, the BER requires approximately 4 photo-electrons per bit (6 dB SNR) to maintain a 10^{-3} bit error rate in WGN while it requires more than 1000 photo-electrons per bit (larger than 30 dB SNR) to maintain the same error rate in fading when no atmospheric compensation techniques are considered. It is clear from these plots that to maintain good receiver sensitivity requires some technique to moderate phase coherence length r_0 such that $D/r_0 = 4$. When phase correction is applied in Figure 10.45(a), the compensation of just 6 modes (phase aberrations are compensated up to astigmatism) brings the number of photo-electrons required to maintain the 10^{-3} BER down to 9 (9.5dB SNR). When we consider MRC diversity combining of the received signal, using eight independent branches (Figure 10.45(b)) also reduces the power to acceptable levels (about 12 photo-electrons per bit, less than 11 dB SNR).

Figure 10.46 is a set of error probabilities from Belmonte and Kahn's analysis that compares modulation formats PSK, QAM, PAM, and DPSK performances. They analyze coherent M-PSK for different constellation sizes ($M = 2, 4, 8, 16$; they noted that theoretically 2- and 4-PSK have similar BER performance). For modulations QAM

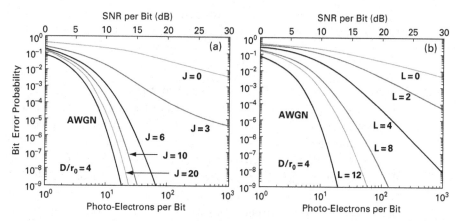

Figure 10.45 Bit-error probability versus turbulence-free photo-electrons ($\eta = 1$) per bit γ_0 for BPSK under coherent detection and additive white Gaussian noise (AWGN) conditions. Performance is shown for different values of: (a) the number of modes J corrected by adaptive optics, and (b) the number of branches L in the combiner.

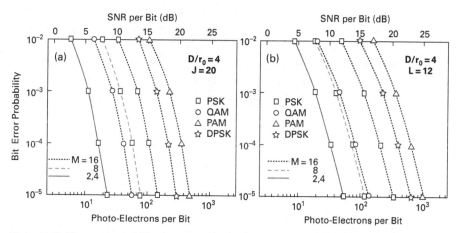

Figure 10.46 Bit-error probability versus turbulence-free photo-electrons ($\eta = 1$) per bit γ_0 for various modulation formats and coded bits per symbol (constellation size M). Coherent detection and additive white Gaussian noise (AWGN) are assumed. Modulation type is indicated by marker style. Number of bits per symbol M is indicated by color. Performance is shown for: (a) phase compensation up to the 5th-order aberrations (J=20), and (b) a set of L=12 branches in the combiner.

PAM, and DPSK, they just considered the high-order 16-ary formats. In addition, they assume 4-QAM is equivalent to 4-PSK. In all modulations analyzed, they have assumed that either turbulence-induced phase aberrations have been compensated up to $J = 20$ (fifth-order aberrations) (Fig. 10.46(a)), or a 12-branch MRC combining has been used (Fig. 10.46(b)). The plots illustrate the photon efficiency for the various binary and non-binary modulation formats using coherent detection.

If the channel bandwidth in Hz is B, and the number of points in the signal constellation is M, Belmonte and Kahn defined the spectral efficiency as

$$S = \frac{R_b}{B} = \frac{R_s R_c \log_2(M)}{B} \tag{10.86}$$

where R_b is the bit rate in bits/s, R_s is the symbol rate in symbols/s, and unit-less parameter $R_c \leq 1$ is the rate of an error-correction encoder that is used to add redundancy to the signal in order to improve the photon efficiency. The uncoded modulations considered in their analysis correspond to $R_c = 1$. In all situations, prevention of inter-symbol interference requires $R_s \leq B$. (Without loss of generality, they assumed the ideal case where $R_s = B$, $1/B$ is the symbol period, and used the number of coded bits per symbol [$\log_2(M)$] as the figure of merit for spectral efficiency.) Figure 10.47 compares spectral efficiency $\log_2(M)$ and photo-electrons per bit requirements for 10^{-9} BER. It clearly illustrates the trade-off between spectral efficiency and photon efficiency of various binary and non-binary modulation formats applied to coherent laser communication through the turbulent atmosphere. Figure 10.47 shows their trade study and lets us look at what they found.

Based on these plots, at spectral efficiencies $S = \log_2(M)$ below 1 bit/s/Hz, 2-PAM and 2-DPSK are attractive techniques. They are simple to implement and lead to the best photo-count efficiencies in terms of photo-electrons per bit required for 10^{-9} BER. Between 1 and 2 bits/s/Hz, 4-DPSK and 4-PSK are perhaps the most interesting modulation formats. At spectral efficiencies above 2 bits/s/Hz, 8-PSK and 16-QAM become the most appealing modulations. In general, the performances of 8-QAM and 8-PSK are very similar because the mean energy of the constellation is just slightly different in both modulations. The complicating factor is that the 8-QAM points are no longer all the same amplitude and so the demodulator must now correctly detect phase and

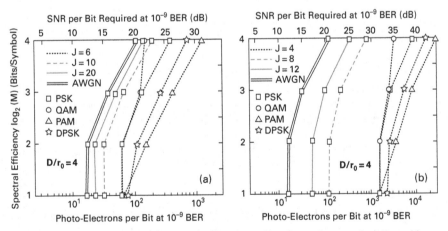

Figure 10.47 Spectral efficiency $\log_2(M)$ vs. turbulence-free photo-electrons ($\eta = 1$) per bit $\gamma_0/\log_2(M)$ requirement for various modulation formats with coherent detection and additive white Gaussian noise (AWGN). Modulation type is indicated by marker style. Performance is shown for different values of: (a) the number of modes J corrected by adaptive optics, and (b) the number of branches L in the combiner.

amplitude, rather than just phase, and consequently 8-PSK is a better choice. For data-rates beyond those of 8-PSK and 8-QAM, it is better to consider QAM since it attains a greater distance between contiguous points in the I-Q plane by distributing the points more squarely. In Figure 10.46(a), the 150 photo-electrons-per-bit required for 16-QAM with phase compensation up to astigmatism ($J = 6$) is better than the near 400-photons-per-bit requirement for the equivalent 16-PSK, and is even better than the near 200-photons-per-bit required for 16-PSK with a higher phase compensation ($J = 10$). With $J = 6$, 16-QAM offers a benefit of more than 4 dB over 16-PSK.

Based on their analysis, Belmonte and Kahn concluded that high communication rates require receivers with good sensitivity along with some technique to mitigate the effect of atmospheric fading. To date, coherent systems have worked best when they are inter-satellite links, but degrade when part of the link is in the atmosphere. For example, Lange *et al.* reported inter-satellite links achieving a BER of 10^{-11} for LEO-LEO links [69]. They also have verified both LEO-to-ground and ground-to-LEO links [70] using 60 mm apertures on ground. Specifically, they have observed some minor burst errors in LEO-to-ground links, but they are error-free for most of the time. In case of ground-to-LEO links, the have achieved a BER of 10^{-5}. When these researchers operated a horizontal link on the Canary Islands, reduced performance was the result. Figure 10.48 shows the bit error rate and eye-diagram for a 142 km, 5.625 Gbps homodyne BPSK FSOC links [63]. The bit error rate varies from 10^{-4} to 10^{-6}. These measurements were taken under medium atmospheric conditions where the Fried parameter r_0 was around 5 cm and link attenuation was >20 dB. The researcher noted that the intensity fluctuations were so strong that they still experienced link disruption even though the phase remained locked. Their system did not compensate for the channel fading to yield error-free communications.

As noted above, countering the effects of the turbulence on the signals is mandatory to obtain the advantages of coherent detection schemes. This is a hot topic in the field today. Approaches using four-wave mixing, coherent fiber arrays and other techniques have been proposed to mitigate the high turbulence in a horizontal FSOC link. For example, Li and his collaborators have proposed new techniques for wavefront correction for these horizontal atmospheric links using coherent arrayed receivers [62,64]. In particular, Figure 10.49 illustrates their basic scheme for DSP-based phase management feed-

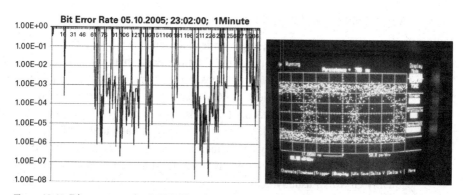

Figure 10.48 Bit error rate for 5.525 Gbps homodyne BPSK FSOC link and eye-diagram [63].

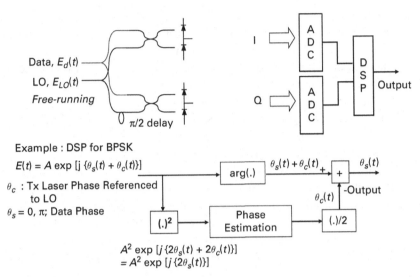

Figure 10.49 DSP-based phase management scheme: feed-forward carrier phase estimation [64].

forward carrier phase estimation. Future experimental validation is needed to determine whether techniques like this, or the other coherent ones, can mitigate this type of negative effect, the decrease in heterodyne efficiency, and provide robust communications as the incoherent systems do.

10.10 Pointing, acquisition and tracking

In Chapter 5, we had a discussion about pointing and tracking. In this section, we will provide more details of a typical pointing, acquisition and tracking (PAT) system for atmospheric FSOC airborne (moving) network. Ground and space optical FSOC systems will operate essentially the same. The intent is to give the reader a basic overview of how an FSOC system gets two or more nodes to talk to one another.

Basic overview of the system

For many FSOC systems, their optical terminal, e.g., MX-15 Skyball, contains an adaptive optics assembly (including the tip/tilt/deformable mirror mechanism, a short-wave infrared (SWIR) wave front sensor and fiber interface), a visible to near infrared (VNIR) narrow field of view (NFOV) camera, a VNIR wide field of view (WFOV) camera and a beacon telescope. (This also is essentially true for both the ORCA and FOENEX systems.) Coupled to this terminal is a navigation-grade (<0.01 degree per hour) global positioning system / inertial navigation system (GPS/INS) for precision pointing, e.g., a Northrop Grumman LN-251. Alignment between the WFOV and high-quality GPS/INS usually is measured after the sensors have been mounted into the node platforms, e.g., shelter and aircraft. Alignment between the WFOV and NFOV cameras usually is measured after the sensors have been assembled. Boresight alignment between the NFOV camera and wave front sensor (WFS) usually are obtained on the ground and *in situ*. In addition to the above, a low-bandwidth RF communications system ("order-wire") is necessary to relay each node's position and velocity to the other nodes.

Figure 10.50 depicts an example PAT state transition diagram for the link acquisition process [71]. In this design, the system sequentially uses the four sources of remote platform positioning: geo-pointing, WFOV sensor, NFOV sensor and WFS in that order, to lock the system into the communications mode. Each step provides greater accuracy in steering the beam to the WFS. The "orderwire" system provides each node with the basic tracking data for acquisition, e.g., latitude, longitude, altitude, velocity, of each of the nodes in the network. Let us assume that the network manager, e.g., IPCM, tell two specific nodes to begin the acquisition process. Following the diagram left to right in Figure 10.50, each node will use the GPS/INS to point their respective clear apertures in the optical terminal to the sector in space where the other node is based. This action uses the latest geo-pointing data provided by the orderwire. Each node will then turn on their beacon and a search in this sector will be conducted by the WFOV camera for the beacon. If the beacon is within the FOV of the WFOV camera (a few 10s of milliradians in angular width), its tracking window will search for the hot spot and, once found, tracks this high-contrast signal, keeping the beacon within the pre-set tracking window. If the beacon is not within the FOV of the WFOV camera, the camera FOV will angularly spiral search from this the initial pointing angle until it finds the beacon, or gets a better geo-pointing update. It should be noted that the beacon can be a CW light with centroid tracking at the

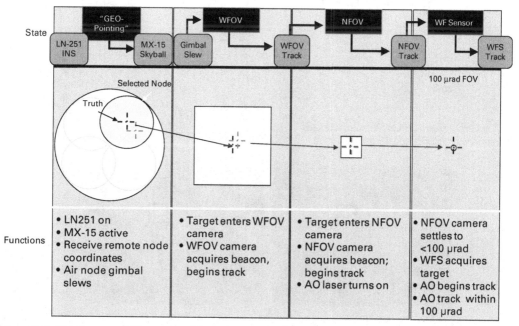

Figure 10.50 Pointing, acquisition and tracking sequence for FSOC applications [70].

receiver or sine-wave-modulated for phase-lock loop tracking the receiver, or whatever scheme the designer preferred.

Example 10.4 The simplest form of tracking is centroid tracking. This is where one tracks either an unresolved or extended object against simple background. Reference [27] provides an excellent overview of this technique and it will be summarized here.

Figure 10.51 shows two realizations of NVIR beacon signal against a blurred background caused by significant turbulence [36,72]. The background clearly is unresolved, blending all clutter into slowly varying background scene, i.e., low-frequency spatial content. Although the signal is unresolved and dynamic, it is of finite spatial extent and of higher intensity than the background, which makes it stand out. This a perfect image for centroid tracking.

As noted above, the basic premise of centroid tracking is localizing a signal profile when it is unresolved, or can be segmented from the background clutter by intensity information alone. To do this, one only looks at a finite number of pixels, usually segmented within a rectangle, or "gate". An example gate is shown in each of the pictures contained in Figure 10.51. The procedure is to essentially segment the object and place a rectangular box, or gate, around it. The gate can be fixed or dynamic in size, e.g., constant false-alarm rate analysis. The size of the gate is chosen to be big enough to cover the possible physical extent of the signal of interest, but not too large as to allow a significant amount of irrelevant background light to influence the algorithm's calculation. Figure 10.51 illustrates two sample stochastic signals of a beacon laser in turbulence, with the gate size chosen to reasonably encompass the unresolved (point) signal.

The basic algorithm for the geometric centroid of an extended/unresolved signal profile is

$$cx = \frac{\sum_{i=1}^{m}\sum_{j=1}^{n} x_i s(i,j)}{S_T} \tag{10.87}$$

and

$$cy = \frac{\sum_{i=1}^{m}\sum_{j=1}^{n} y_i s(i,j)}{S_T} \tag{10.88}$$

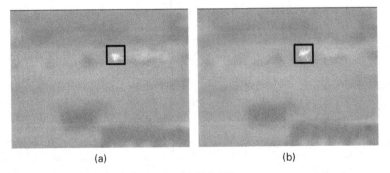

(a)   (b)

Figure 10.51 Centroid tracking examples [36, 73].

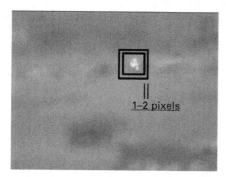

Figure 10.52 Gate for calculating local image statistics.

where

$$S_T = \sum_{i=1}^{m} \sum_{j=1}^{n} s(i, j) \tag{10.89}$$

[27]. The indices (i, j) represent the pixel locations in the x and y direction, respectively, $s(i, j)$ is the image intensity at (i, j), and S_T is the total intensity in the gate. One computes the centroid inside the gate for the entire image and then compares it to a threshold, or picks the largest. Here, the centroid does not track a specific feature of the signal, rather determines the location of the maximum signal energy.[4]

The signal-to-noise ratio is given by

$$\text{Signal\_SNR} = \frac{\sum_{target} \sum_{target} s(i, j)}{\sigma_{Noise}} \tag{10.90}$$

where σ_{Noise} is the standard deviation of the local image noise. Typically, the mean noise intensity and noise standard deviation for the local image area are calculated using a rectangular gate, 1–2 pixels wide, surrounding the target area, e.g., around the tracking gate [71–73]. An example of this statistics gate is depicted in Figure 10.52.

When an intensity threshold is used, there are three first-order methods to choose from: binary, Type I and Type II [27]. The key characteristics of these types are [27]

- Binary
  - Pixels above threshold are treated as "1"s and those below as "0"s.
  - It is an incentive to intensify variations of the signal (which obviously occurs in Figure 10.51).
  - It is usually used for extended targets.
- Intensity Type I
  - Signal above threshold retains value, below is set to zero [73,75].
- Intensity Type II
  - Subtract threshold from signal above threshold. Pixels below threshold value are set to zero [73,75].

---

[4] This discussion is valid for extended target tracking as well as for beacon type signals.

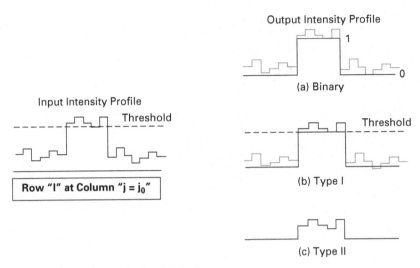

Figure 10.53 Three example threshold schemes.

Figure 10.53 illustrates these methods graphically on a row of pixel intensities. They can be applied to either positive or negative contrast signals, with respect to the background. Other more complicated schemes can be used, employing two or more thresholds. These are useful in cases where the target contrast is bimodal (e.g., aircraft against a sky background, beacon against ground clutter). This is sometimes referred to as "exterior" threshold where high-intensity and low-intensity pixels are passed while midrange pixels are excluded [27]. Another case, called "interior" threshold, occurs when the target is in the midranges and the background is both brighter and darker than the target. This occurs in the IR when a tank crosses a paved road.

The following are centroid tracking positives and negatives [27]:

- Good where signal/target can be segmented from background based on intensity differences.
- It can be shown to be "optimal" in least-squares sense for parabolic intensity distribution.
- Binary is good for extended targets where glints may be a problem.
- Type I leads to better noise performance than Type II or binary.
- When quantization effects are considered Type II is best, binary is worst.

Once the WFOV tracking is stable and locked, the signal location will be handed off (angle-angle command) to the NFOV camera. If the beacon is within the FOV of the NFOV camera (a few milliradians in angular width), its tracking window will search for the hot spot and, once found, tracks this high-contrast signal, keeping the beacon within the pre-set tracking window. If not found, the NFOV camera system hands the search back to the WFOV system and the previous process repeats.

Once the NFOV tracking is stable and settles to less than a 100 μrad pointing error, the communications laser will be turned on and the NFOV signal location will be handed off to the WFS. The WFS acquires and tracks the signal (PIB) in 100 μrad. This leads to the AO system assuming the tracking function, still within the 100 μrad. If the signal is lost, the WFS transfers the search back to the NFOV camera and the previous process repeats. (In both FOENEX and ORCA, the PAT systems worked so well that the WFOV camera typically sent the angle-angle command directly to the WFS, circumventing the NFOV step.)

When a loss in link connectivity occurs (e.g., due to a cloud or wing blockage), the tracking error will grow as the position of the nodes becomes more uncertain. Typically, a multi-state-state extended Kalman filter (EKF) will be used to predict the trajectory of each node until new navigation data are provided via the discovery system. As the error grows in time, given no new navigation data arrive, the likelihood that the node is within the sensors' field of view is decreased. (For example, a six-state EKF only has position and velocity accounted for. Without the acceleration state vector, one cannot project accurately whether a speed and/or turn will occur during the outage if it occurs.) If the optical terminal seeks to maintain the link with the remote aircraft, the time required to re-establish the link is dependent upon the magnitude of the pointing error. If the pointing error remains within the FOVs of the WFOV or NFOV cameras, then the link can be re-established relatively quickly (milliseconds). If the error grows beyond the FOV of the WFOV camera, then the system either begins the spiral search, or goes back to the geo-pointing state of Figure 10.50 in order to get updated navigational messages to re-establish the link.

Example 10.5 This example looks at the resulting tracking error of the airplane-based ATP for a six-state EKF under extended disruptions of the link and navigational data [76].

We begin with a derivation of the pointing error model. This development is based on the diagram in Figure 10.54 and uses basic geometric relationships [76]. It estimates the tracking error as a function of aircraft speed, turn rate and outage time. When a loss of a hybrid link, beacon, and navigational updates occurs, we want to predict the expected position along the straight trajectory at some distance d, but the actual position is along the turn trajectory.

Referring to Figure 10.54, the maximum offset between the predicted and actual location of the remote aircraft is given by:

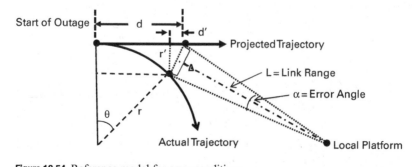

Figure 10.54 Reference model for error conditions.

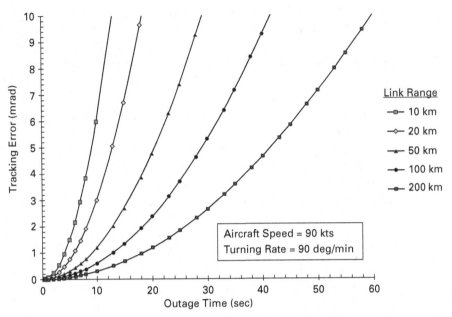

Figure 10.55 Tracking error angle as a function of outage time for aircraft flying at 90 knots and with a turning rate of 90 degrees per minute.

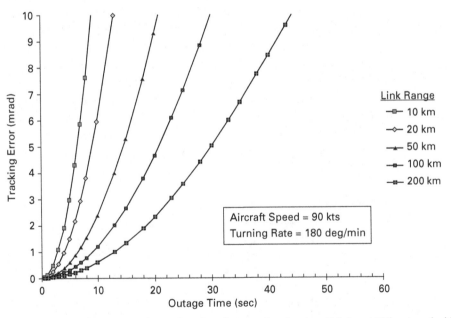

Figure 10.56 Tracking error angle as a function of outage time for aircraft flying at 90 knots and with a turning rate of 180 degrees per minute.

$$\Delta = \sqrt{r'^2 + d'^2} \tag{10.91}$$

The segments r' and d' are found using the equations:

$$r' = r(1 - \cos\theta) \tag{10.92}$$

$$d' = d - r\sin\theta \tag{10.93}$$

respectively. Assuming an aircraft flying at speed v in a steady turn with turn rate $\theta^{\cdot}$, the turn radius r is approximately given by $\theta^{\cdot}$. The expected position d at time t is located along the straight trajectory at a distance of vt from the last tracking update and the turn angle θ is $\theta^{\cdot} t$. The angular tracking error relative to the expected position is a function of range L and error Δ and is given by the equation

$$\alpha = \arctan\left(\frac{\Delta}{2L}\right) \tag{10.94}$$

Figures 10.55 and 10.56 depict the tracking error angle as a function of outage time for aircraft flying at 90 knots and with a turning rate of 90 and 180 degrees per minute, respectively, and a fixed node. Figure 10.57 depicts the tracking error angle as a function of outage time for aircraft flying at 180 knots and with a turning rate of 90 degrees per minute and a fixed node. Figure 10.55 shows that the larger the distance between the nodes, the slower the tracking error angle grows. This is because the angular separation

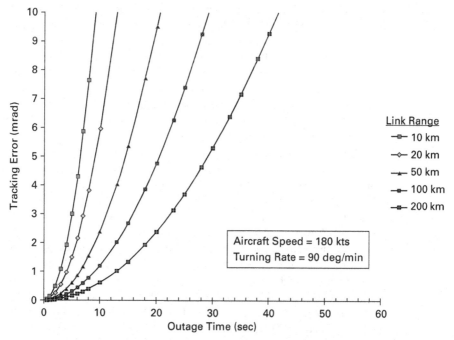

Figure 10.57 Tracking error angle as a function of outage time for aircraft flying at 180 knots and with a turning rate of 90 degrees per minute.

between the curved and straight trajectory gets smaller and it takes longer in time for a larger separation to occur with either a speed or turn rate change at the longer distances. Comparing Figures 10.55 and 10.56, we see that the faster the turning rate, the faster the tracking error grows. Here the faster turning rate increases the tracking error faster, even at the longer separations. Comparing Figures 10.55 and 10.57, we also see that the faster the aircraft speed, the faster tracking error grows. Speed also creates faster error growth, even at the larger separations. Comparing all three figures, we see the tracking error grows slightly faster for the higher speed than for the higher turning rate.

References

1. A. Majumdar and J. Ricklin. *Free Space Laser Communications: Principles and Advances*, Optical and Fiber Communications Series. Springer Science+Business Media, New York (2008).
2. L. C. Andrews and R. L. Phillips. *Laser Beam Propagation through Random Media*, 2nd edn. SPIE Press (2005).
3. L. C. Andrews. *Field Guide to Atmospheric Optics*. SPIE Press (2004).
4. R. R. Beland. *Propagation Through Atmospheric Optical Turbulence,* Ch. 2 in Vol. 2 of *Infrared and Electro-Optical Systems Handbook*. Environmental Research Institute of Michigan (1996).
5. S. Karp, R. M. Gagliardi, S. E. Moran and L. B. Stotts. *Optical Channels: Fiber, Atmosphere, Water and Clouds*. Plenum, New York (1988), Ch. 5.
6. W. L. Wolfe and G. J. Zissis. *The Infrared Handbook*. IRIA Center, ERIM (1993).
7. D. Fried. For a coherent laser radar: the power spectral density of the detected signal's phase error due to turbulence effects, expressed in terms of an equivalent laser's frequency jitter power spectral density. Report No. TN 220R (March 2007).
8. L. C. Andrews, R. L. Phillips, R. Crabbs, D. Wayne, T. Leclerc and P. Sauer. Atmospheric channel characterization for ORCA testing at NTTR. Atmospheric and oceanic propagation of electromagnetic waves IV. *Proceedings of SPIE*, **7588** (2010).
9. D. L. Walters and K. E. Kunkel. Atmospheric modulation transfer function for desert and mountain locations: the atmospheric effects on r_0. *Journal of the Optical Society of America*, **71** (1981), pp. 397–405.
10. L. C. Andrews, R. L. Phillips, D. Wayne *et al.* Near-ground vertical profile of refractive-index fluctuations. *Proceedings of SPIE*, **7324** (2009).
11. M. Born and E. Wolf. *Principles of Optics: Electromagnetic Theory of Propagation, Interference and Diffraction of Light*. Cambridge University Press, London (1999).
12. K. Strehl. *Zeitschrift für Instrumentkunde*, **22** (1902), p. 213.
13. V. N. Mahajan. Strehl ratio for primary aberrations: some analytical results for circular and annular pupils. *Journal of the Optical Society of America*, **72** (1982), pp. 1258–1266.
14. V. N. Mahajan. Strehl ratio for primary aberrations: some analytical results for circular and annular pupils. *Journal of the Optical Society of America*, **72** (1982), pp. 1258–1266. Strehl ratio for primary aberrations in terms of their aberration variance. *Journal of the Optical Society of America*, **73** (1983), pp. 860–861.
15. D. L. Fried. Optical heterodyne detection of an atmospherically distorted signal wavefront. *Proceedings of the IEEE*, **55** (1967), pp. 57–67.

16. R. K. Tyson. *Field Guide to Adaptive Optics*. SPIE Field Guides, Volume FG03, John E. Greivenkamp, Ed. SPIE Press, Bellingham, WA (2004).

17. R. K. Tyson. *Introduction to Adaptive Optics*, Tutorial Texts in Optical Engineering, Vol. TT41, Arthur R. Weeks, Jr, Ed. SPIE Press, Bellingham, WA (2000).

18. Michael C. Roggemann and Byron Welsh. *Imaging through Turbulence*. CRC Press, Boca Raton, FL (1996).

19. J. W. Hardy. Adaptive optics. *Scientific American*, **260**(6) (June 1994), pp. 60–65.

20. R. K. Tyson. *Principles of Adaptive Optics*, 2nd edn. Academic Press (1991).

21. J. W. Hardy. Adaptive Optics – a new technology for the control of light. *Proceedings of the IEEE*, **66** (1978), pp. 651–697.

22. L. B. Stotts, B. Stadler, B. Graves *et al*. Optical RF Communications Adjunct. *Proceedings of the SPIE Conference on Free-Space Laser Communications VIII*, Arun K. Majumdar and Christopher Davis, Eds. Volume 7091, 10–12 August 2008.

23. Anthony Seward. M Z A, from www.mza.com/publications/MZADEPSBCSMSCP0.ppt.

24. L. C. Andrews and R. L. Phillips. University of Central Florida, private communications.

25. H. Alan Pike. private communication. See Lord Rayleigh. *Scientific Papers*, Vol I, p. 491. Cambridge University Press, London (1899–1920).

26. L. B. Stotts, J. Foshee, B. Stadler *et al*. Hybrid optical RF communications. *Proceedings of IEEE Conference*, **97**(6) (2009), pp. 1109–1127.

27. ATP Section 4. Image Processing, Tracking Algorithms and. Presentation Summary: IV-1, Class IV. Spring 2009: http://www.davincisworld.com/Light/SystemPerfTools/ATP_Section_4_Image_Processing_Algorithms.ppt.

28. F. Roddier. Maximum gain and efficiency of adaptive optics systems. *Publications of the Astronomical Society of the Pacific*, **109** (1998), pp. 815–820.

29. L. B. Stotts. Optical RF communications adjunct. AFCEA/IEEE Military Communications (MILCOM) 2008: Lasercom as an enabling technology for high bandwidth communications. San Diego, CA, 17 November 2008.

30. L. B. Stotts, B. Stadler, D. Hughes *et al*. Optical communications in atmospheric turbulence. SPIE Free Space Laser Communications Conference (2009).

31. R. J. Noll. Zernike polynomials and atmospheric turbulence. *Journal of the Optical Society of America*, **66**(3) (1976), pp. 207–211.

32. L. C. Andrews. University of Central Florida, private communications.

33. J. C. Juarez, D. W. Young , R. A. Venkat *et al*. Analysis of link performance for the FOENEX laser communications system. Defense Sensing & Security Symposium 2012, Laser Sensors and Systems, Conference: Atmospheric Propagation IX, Baltimore, MD, 25–26 April 2012. *Proceedings of the SPIE*, **8380**, Linda M. Wasiczko Thomas, U.S. Naval Research Lab.; Earl J. Spillar, US Air Force (2012).

34. H. T. Friis. *Proceedings of the IRE*, **34** (1946), p. 254.

35. H. A. Pike, L. B. Stotts, P. Kolodzy and M. Northcott. Parameter estimates for free space optical communication. Application of lasers for sensing & free space communication (LS&C). Optical Society of America, Toronto, Ontario, Canada, 10–14 July 2011.

36. L. B. Stotts, P. Kolodzy, H. A. Pike *et al*. Free space optical communications link budget estimation. *Applied Optics*, **49**(28) (2010), pp. 5333–5343.

37. R. S. Kennedy and S. Karp, eds. Optical Space Communications, NASA SP-217, Appendix G (D. Fried) (1969), pp. 135–138.

38. J. H. Shapiro. Reciprocity of the turbulent atmosphere. *Journal of the Optical Society of America*, **61** (1971), pp. 492–495.

39. R. G. Walther. Diversity in air-to-ground lasercom: the FOCAL Demonstration. Technical Panel, Session DoD-2: Freespace Optical Communications. 2011 Military Communications Conference (MILCOM 2011), Baltimore, MD, 7–10 November 2011.

40. R. R. Parenti, J. M. Roth, J. H. Shapiro and F. G. Walther. Observations of channel reciprocity in optical free-space communications experiments. OSA Conference on Applications of Lasers for Sensing & Free Space Communications (2011).

41. F. G. Walther, S. Michael, R. R. Parenti and J. A. Taylor. Air-to-ground lasercom system demonstration design overview and results summary. *Proceedings of SPIE*, **7814**, Free-Space Laser Communications X (2010).

42. S. Michael, F. G. Walther and R. R. Parenti. Performance evaluation of an air-to-ground optical communications demonstration. OSA LS&C meeting, February 2010.

43. R. R. Parenti, S. Michael, J. M. Roth and T. M. Yarnall. Comparisons of C_n^2 measurements and power-in-fiber data from two long-path free-space optical communications experiments. *Proceedings of SPIE*, **7814**, Free-Space Laser Communications X (2010).

44. J. A. Greco. Design of the high-speed framing, FEC, and interleaving hardware used in a 5.4km free-space laser communications experiment. *Proceedings of SPIE*, **7464**, Free-Space Laser Communications IX (2009).

45. S. Michael *et al.* The use of statistical channel models, full-field propagation codes, and field data to predict link availability. *Proceedings of SPIE*, **7464**, Free-Space Laser Communications IX (2009).

46. J. Moores F. G. Walther J. A. Greco *et al.* Architecture overview and data summary of a 5.4km free-space laser communications experiment. *Proceedings of SPIE*, **7464**, Free-Space Laser Communications IX (2009).

47. R. J. Murphy *et al.* A conical scan free space optical tracking system for fading channels. *Proceedings of SPIE*, **7464**, Free-Space Laser Communications IX (2009).

48. F. G. Walther *et al.* A process for free-space laser communications system design. *Proceedings of SPIE*, **7464**, Free-Space Laser Communications IX (2009).

49. T. Williams *et al.* A free-space optical terminal for fading channels. *Proceedings of SPIE*, **7464**, Free-Space Laser Communications IX (2009).

50. S. Michael *et al.* Comparison of scintillation measurements from a 5 km communications link to standard statistical models. *Atmospheric Propagation VI*, Linda M. Wasiczko Thomas & G. Charmaine Gilbreath, eds. SPIE, Orlando, FL (2009).

51. F. G. Walther, G. A. Nowak, S. Michael *et al.* Air to ground lasercom systems demonstration. *Proceeding of the 2010 Military Communications Conference (MILCOM 2010)*, San Jose, California, USA, 31 October–3 November 2010.

52. M. Fletcher, J. Cunningham, D. Baber *et al.* Observations of atmospheric effects for FALCON laser communication system Flight test. Atmospheric Propagation VIII. *Proceedings of SPIE*, **8038** (2011), Paper F.

53. J. Cunningham, D. Foulke, T. Goode *et al.* Long range field testing of free space optical communications terminals on mobile platforms, radio and optical communications (U112). MILCOM 2009, paper 901469, 20 October 2009.

54. D. Young, J. Sluz, J. Juarez *et al.* Demonstration of high data rate wavelength division multiplexed transmission over a 150 km free space optical link. MILCOM 2007, Advanced Communications Technologies 4.2, Directional Hybrid Optical/RF Networks (2007).

55. J. C. Juarez, D. W. Young and J. E. Sluz. Optical automatic gain controller for high-bandwidth free-space optical communication links. Application of Lasers for Sensing & Free Space Communication (LS&C), Optical Society of America, Toronto, Ontario, Canada, 10–14 July 2011.

56. J. C. Juarez, D. W. Young, J. E. Sluz and L. B. Stotts. High-sensitivity DPSK receiver for high-bandwidth free-space optical communication links. *Optics Express*, **19**(11) (2011), pp. 10789–10796.

57. L. B. Stotts, N. D. Plasson, T. W. Martin, D. W. Young and J. C. Juarez. Progress towards reliable free-space optical network. MILCOM 2011, Baltimore, MD, 7–10 November 2011.

58. V. Vilnrotter and M. Srinviasan. Adaptive detector arrays for optical communications receivers. *IEEE Transactions on Communications*, **50** (2002), pp. 1091–1097.

59. I. Kim, G. Goldfarb and G. Li. Electronic wavefront correction for PSK free-space optical communications. *Electronic Letters*, **43**(20) (27 September 2007).

60. M. Fernandez and V. Vilnrotter. Coherent optical receiver for PPM signals received through atmospheric turbulence: performance analysis and preliminary experimental results. *Proceedings of SPIE*, **5338** (2004), pp. 151–162.

61. K. Kikuchi. Coherent detection of phase-shift keying signals using digital carrier-phase estimation. OFC, OTuI4, Anaheim, CA, 2006.

62. I. Kim, C. Kim and G. Li. Requirements for the sampling source in coherent linear sampling. *Optics Express*, **12**(12) (2004), pp. 2723–2730.

63. R. Lange, B. Smutny, B. Wandernoth, R. Czichy and D. Giggenbach. 142 km, 5.625 Gpbs free-space optical link based on homodyne BPSK modulation. Free-Space Optical Communication Technologies XVIII, edited by G. Stephen Mecherle. *Proceedings Of SPIE*, **6105A** (2006), pp. 1–9.

64. G. Li. Coherent optical technologies for free-space optical communication and sensing. Application of Lasers for Sensing & Free Space Communication (LS&C), Optical Society of America, Toronto, Ontario, Canada, 10–14 July 2011.

65. A. Belmonte and J. M. Kahn. Efficiency of complex modulation methods in coherent free-space optical links. *Optics Express*, **18**(4) (2010), pp. 3928–3837.

66. A. Belmonte and J. Khan. Performance of synchronous optical receivers using atmospheric compensation techniques. *Optics Express*, **16**(18) (2008), pp. 14151–14162.

67. A. Belmonte and J. M. Kahn. Capacity of coherent free-space optical links using diversity-combining techniques. *Optics Express*, **17**(15) (2009), pp. 12601–12611.

68. J. W. Goodman. *Speckle Phenomena in Optics. Theory and Applications*. Ben Roberts & Company, New York (2007).

69. J. D. Parsons. Diversity techniques in communications receivers. In , D. A. Creasey, ed., *Advanced Signal Processing*. Peregrines (1985), Ch. 6.

70. F. Heine, H. Kämpfner, R. Lange *et al*. Laser communication applied for EDRS, the European data relay system. *CEAS Space Journal*, **2** (2011), pp. 85–90.

71. R. Lange. Private communications.

72. Courtesy of David Abelson and colleagues, AOptix

73. Courtesy of Dr. David Wayne, University of Central Florida.

74. L. B. Stotts, E. M. Winter, L. E. Hoff, and I. S. Reed. Clutter rejection using multi-spectral processing. *Proceedings of SPIE*, **1305** (1990).

75. I. S. Reed, L. B. Stotts and R. M. Gagliardi. A recursive moving target indication algorithm. *IEEE Transactions on Aerospace and Electronic Systems*, **26**(3) (1990), p. 434.

76. I. S. Reed, L. B. Stotts and R. M. Gagliardi. Optical moving target detection with three-dimensional matched filtering. *IEEE Transactions on Aerospace and Electronic Systems*, **24**(4) (1988), p. 327.

77. T. Martin, Science and Technology Associates. Private communications.

11 Communications in the optical scatter channel

11.1 Introduction

In Chapter 10, we discussed the effects on high-data-rate FSOC systems from atmospheric turbulence, which are dominated by scintillation / channel fading and beam wander, how they are modeled and how they are mitigated to yield high communications link / network performance. From some of the figures there, one can see channel dynamics on a very fast scale. When clouds insert themselves in the link, we suggested that RF communications (a hybrid system) could provide a means for keeping the link connected at a reasonably high, but much lower, data rate. This works well in the atmosphere in those climates where clouds are infrequent or sparse. This strategy does not work in the optical scatter channel where particulate absorption and scattering significantly degrade the incoming signal to the point that the original diffraction-limited beam becomes lost in the system noise floor. Alternate strategies must be employed to facilitate high communications link / network availability because of the significant degradation of the original signal by the atmospheric and maritime optical scatter channels. The signal structures that result from each channel are quite different from each other, as well as significantly different from the turbulence channel. Kennedy was one of the first to recognize that these significant differences in structure from the former relative to the latter could not be easily mitigated; he suggested that optical system designers exploit their new structures in order to close the communications link as typical mitigation techniques were useless in the diffusive scattering regime that defines high link availability [1]. Chapters 5 and 9 showed that it could be used for target imaging. This chapter will discuss how this approach can be used for communications in the optical scattering channel, highlighting the models and techniques used by today's researchers and engineers.

11.2 Optical scatter channel models

Chapter 5 introduced the optical scatter channel by describing Mie scattering and its effect on optical signals. This introduced the inherent properties of the optical channel. In this section, we will expand this discussion, focusing on the detailed effects on laser communications created by the atmospheric and maritime scatter channels, and their modeling. Each has its own unique characteristics.

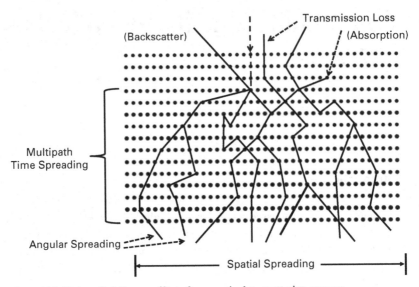

Figure 11.1 Various light beam effects from particulate scattering process.

The atmospheric optical scatter channel has negligible absorption (single scatter albedo $\omega_0 = b/c$, greater than 0.99), dispersing the signal in space, time and angle. This is illustrated in Figure 11.1. Ironically, on the stormiest day, the maximum solar irradiance loss is essentially only 10 dB or so. That is the good news. Unfortunately, the radiance emerging from the cloud is Lambertian in nature, increasing the rms-beam angle of around a quarter-degree to 10s of degrees. In addition, the time dispersion of any narrow temporal signal increases to the 10s to 100s of microseconds. That is the bad news. These latter effects allow external background noise like solar light to be received by the optical detector, enforcing background-noise-limited communications during the day (night-time operation is usually quantum-limited communications).

On the other hand, the maritime optical scatter has significant absorption, which limits the amount of dispersion of the signal in space, time and angle, but the signal degradation is still significant. Unlike a cloud layer that exhibits uniform sunlight illumination across its surface, i.e., the profile of the Sun is lost, the maritime optical channel has a "blurred sun" always overhead. This is known as the manhole effect. To first order, this "sun" has a Gaussian radiance distribution, always centered on the up-down axis that decreases in depth exponentially. Although large-angle scattering is dispersing the incoming light within the channel, this scattered light also has an increased probability of being absorbed in future interactions with particles before it hits the optical receiver aperture. The result is that only light that is scattered a few times, mostly down, makes it to the optical receiver aperture with the aforementioned radiance distribution.

11.2.1   Analytical models and relationships for optical propagation through clouds

As noted in Chapter 4, optimum performance depends upon both the total power and the form of the received radiance after passing through the optical scatter channel. Above, we provided a general overview of the effects on light beams by clouds and the water column, the two primary components of this channel. This section will describe a set of engineering models for optical propagation through clouds. The next section will deal with propagation through water.

The four major effects that light has on optical beams are

- angular spreading
- spatial spreading
- temporal spreading
- transmission loss (including absorption and back scatter).

We illustrate these effects graphically in Figure 11.1. For most system applications requiring high link availability, one must consider clouds with large optical thickness, $\tau = cz$, of the order of 100 or more, and at depths where both regimes are in the diffusion regimes of scattering. Bucher [2,4] and Lee et al. [2,5] provided a very useful set of engineering models for characterizing the above effects for clouds and water using Monte Carlo simulations. These equations are very useful in creating link budgets for the optical communications. The angular distribution of light emerging from the bottom, or top, of a cloud, or from the ocean surface, follows essentially a Lambertian distribution [2, p. 13]. We examined means for calculating the received power from a Lambertian source in Chapter 1. In the next subsections, we will describe these engineering models for the remaining effects that will be used to calculate the resulting loss they create. Example link budgets also will be given to show their use.

11.2.2   Spatial spreading in thick clouds

Spatial spreading is the additional dimensional increase in a beam's finite cross section apart from diffraction. Bucher found that the radius that contains half of the transmitted rays from the medium is given by

$$r_c = 0.92z(1-\tau_d)^{-0.08} \tag{11.1}$$

where

$$\tau_d = \tau(1-g) \tag{11.2}$$

and

$$g = \frac{1}{4\pi} \int_0^{2\pi} \int_0^{\pi} p(\theta, \phi) \cos\theta \, d\Omega \tag{11.3}$$

[2,3]. The parameter τ_d is known as the diffusion thickness. The parameter g is called the asymmetry factor and is the average cosine of the scalar phase function of the medium, p (θ, φ). The scalar phase function was introduced in Chapter 4. The form of Eq. (11.11) implies that the spatial spreading saturates and even decreases as a cloud of fixed physical thickness becomes more optically dense.

11.2.3   Transmission loss in thick clouds

Danielson *et al.* developed an equation for the transmission loss that results from light scattering within thick clouds [2,6]. They showed that transmittance for a non-absorptive thick cloud was inversely dependent on the medium's diffusion thickness. Bucher verified this result through Monte Carlo simulation, and developed an analytical expression for this effect. Specifically, he showed that the transmittance/transmission loss for non-absorptive thick clouds is given by

$$L_c = \frac{A(\phi_s)}{1.41 + \tau_d} \tag{11.4}$$

where

$$A(\phi_s) = 1.69 - 0.5513\,\phi_s + 02.7173\,\phi_s^2 - 6.9866\,\phi_s^3$$
$$+ 7.1446\,\phi_s^4 - 3.4249;\,\phi_s^5 + 0.6155\,\phi_s^6 \tag{11.5}$$

In the above, ϕ_s is the signal/solar zenith angle (relative to the cloud top normal) in radians. Figure 11.2 gives cloud transmittance as a function of optical thickness for an asymmetry factor of 0.875 and various incident zenith angles. From this figure, we see that the transmittance decreases 6 dB from 0 to 90 degree incident angles, and follows $\sqrt{\cos\phi_s}$ for $\phi_s < 85°$.

Lee *et al.* investigated the effect of absorption on the multiple-scattering process [5]. As noted above, they found that angular spreading decreases because of the increased probability of light being absorbed from the additional path incurred from the particulate scattering process. The result is the cloud transmittance in Eq. (11.5) was modified to be

$$L_c = \frac{A(\phi_s)}{1.41 + \tau_d} \frac{2k'(\tau + g_1)\exp[-k'(\tau + g_1)]}{1 - \exp[-2k'(\tau + g_1)]} \tag{11.6}$$

where

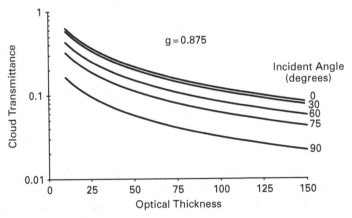

Figure 11.2 Cloud transmittance as a function of optical thickness for various zenith angles.

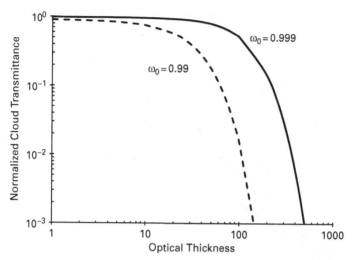

Figure 11.3 Normalized transmittance of absorptive clouds as function of optical thickness for two single scatter albedos.

$$k' = \sqrt{3\,(1-g_1)\,(1-\omega_0)}, \tag{11.7}$$

$$g_1 = \frac{1.42}{(1-g)} \tag{11.8}$$

and

$$\omega_0 = \frac{b}{c} \tag{11.9}$$

Figure 11.3 shows normalized cloud transmittance [ratio of Eq. (11.6) to Eq. (11.4)] as a function of optical thickness for two values of the single scatter albedo, ω_0. It is clear that a little absorption can significantly affect transmission loss for optical thicknesses greater than an optical thickness of 20.

11.2.4 Multipath time spreading in absorptive particulate media

In addition to cloud transmittance, Lee *et al.* investigated the effect of particulate scattering on laser pulse propagation and developed the following equation for multipath time spreading (temporal spreading):

$$\sigma_t = \frac{1}{c_0}\frac{0.58\,D\,\tau_d^{1.5}/\,(1-g)}{0.49 + (1-\omega_0)\,\tau_d^{1.1}/\,(1-g)} \tag{11.10}$$

where c_0 is the speed of light ($=299,792.458$ kilometers per second, or $\sim 3.0 \times 10^8$ m/s) and D is the scattering length of the medium [5]. Figure 11.4 depicts the temporal spreading due to an absorptive cloud as a function of optical thickness for three values of single-scatter albedo. Similarly to the normalized cloud transmittance, the temporal spreading can be affected significantly for optical thickness greater than 20 as the absorption in the cloud increases.

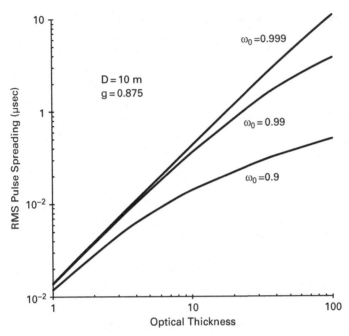

Figure 11.4 RMS Pulse spreading in absorptive clouds as function of optical thickness for various single scatter albedos.

11.3   Analytical models and relationships for optical propagation in the ocean

Recall from Chapter 5 the introduction of the inherent optical scatter properties of a medium like ocean water. The apparent optical properties of a medium are those perceived as general features of the light distribution found with a particulate, multiple-scattering medium [7,8]. They can be developed either through solutions of the radiative transfer equation, or through direct measurements of specific quantities in the field. These properties are most useful in describing situations where absorption is clearly present and the scattering is approaching, or is in, the diffusion regime [9]. Most important, naval systems such as submarine laser communications and underwater mine detection operate under those circumstances. Those situations also pertain to many environmental sensing applications as well. In addition, those situations allow simple, practical answers for a wide range of multiple-scattering problems to be developed, and exact mathematical relationships between themselves and the more fundamental inherent properties to be formulated. For those reasons, this section will focus on the apparent properties of the ocean near, and in, the diffusion regime, where the diffuse attenuation coefficient plays a preeminent role. For those readers interested in more details on the mathematical modeling of shallow-water light propagation, reference books such as *Optical Channels* [2] and papers founds in the Proceedings of SPIE Ocean Optics Conference Proceedings, the Institute of Electronic and Electrical Engineers, Optical Society of America and *Applied Optics* are recommended.

11.3.1   Reflectance of the water column

The reflectance of a water column is defined as the ratio of the upwelling to downwelling radiance at depth z. Mathematically, we write

$$R = \frac{E_u(z)}{E_d(z)} \tag{11.11}$$

where

$$E_d(z) = \int_0^{2\pi} \int_0^{\pi/2} N(z, \theta, \phi) \cos\theta \, d\Omega \tag{11.12}$$

$$E_u(z) = \int_0^{2\pi} \int_{\pi/2}^{\pi} N(z, \theta, \phi) \cos\theta \, d\Omega \tag{11.13}$$

and

$$d\Omega = \sin\theta \, d\theta \, d\phi \tag{11.14}$$

with $N(z, \theta, \phi)$ being the radiance distribution at depth. In these equations, $E_u(Z)$ represents the flux per unit area per unit wavelength interval measured by a horizontally oriented cosine collector facing downward to accept upwelling radiance, and $E_d(Z)$ the reverse [9,10]. R is also known as the irradiance ratio.

For remote sensing purposes, a more useful definition of reflectance is the ratio of upwelling to downwelling irradiance. This definition is called the directional reflectance and is written as

$$\rho = \frac{\pi N_u(z)}{E_u(z)} \tag{11.15}$$

This equation gives the ratio of the radiant flux returned (reflected) by the water column to that which would be reflected into the same solid angle under the same conditions of incidence by a perfect diffuser [11]. Austin *et al.* have shown that the above reflectance is related via the equation

$$R = \frac{E_u(z)}{E_d(z)} = \frac{N_u(z)Q}{E_d(z)} \tag{11.16}$$

$$= \rho Q \Big/ \pi \tag{11.17}$$

where Q is the compensation factor needed to account for the fact that the upwelling radiance distribution may not be uniform at the depth of interest [12]. It is typically between 0.95 and 3.82, and is wavelength-dependent.

11.3.2   Radiance distribution within the water column

The irradiance attenuation coefficients for natural waters are defined as the logarithmic depth derivatives of the upwelling and downwelling irradiance. Mathematically, they are

$$k_u = \frac{-1}{E_u(z)} \frac{d E_u(z)}{dz} \tag{11.18}$$

and

$$k_d = \frac{-1}{E_d(z)} \frac{d E_d(z)}{dz} \tag{11.19}$$

Where k_u and k_d are the upwelling and downwelling irradiance attenuation coefficients, respectively. In practice, these coefficients are usually determined from log-irradiance versus depth plots, $k_{u,d}$ being the slope of the resulting curves.

In an unbounded medium with uniform scattering and absorption properties, the asymptotic radiance distribution for plane wave illumination can be shown to be given by [7]:

$$N_\infty(z, \theta) = S_1 N(u) \exp[-kz] + S_{2_1} N(-u) \exp[kz] \tag{11.20}$$

where $u = \cos\theta$, $S_1 \equiv$ "diffusion stream" in positive z-direction, $S_2 \equiv$ "diffusion stream" in negative z-direction, $k \equiv$ asymptotic radiance attenuation coefficient, and $N(u) \equiv$ diffusion pattern (derived from the medium's scalar phase function and single-scatter albedo. Here, k is the smallest eigenvalue, and $N(u)$ is the corresponding eigenfunction, both derivable from the equation:

$$[1 - ku] N(u) = \int_{-1}^{+1} \int_{-1}^{+1} h(u, v) N(v) \, dv \tag{11.21}$$

with $h(u, v)$ being the redistribution function defined as the average over azimuth of the medium's scalar phase function. In the asymptotic regime, it can be shown that

$$k_d(z) = k = k_u(z)$$

[7]. In optical oceanography, k is called the diffuse attenuation coefficient, which we were exposed to in Chapter 5. It is most useful in optical sensing and communications analysis since Eq. (11.21) gives the exact irradiance/radiance loss rate at deep depths and a conservative estimate of that rate in the diffusive transition region. From thermal dynamic considerations, it lies between the volume absorption coefficient and the volume extinction coefficient, or $a \leq k \leq c$. Let us see how this parameter relates to the inherent properties.

Figure 11.5 depicts the ratio of the diffuse attenuation coefficient to the volume extinction coefficient as a function of the single-scatter albedo for selected asymmetry factors, g. Also shown in this figure are values for representative scattering media. The asymmetry factor is the average cosine of the scalar phase function of the medium, i.e.,

$$g = \frac{1}{4\pi} \int_0^{2\pi} \int_0^\pi P(\theta, \phi) \cos\theta \, d\Omega \tag{11.22}$$

For small-angle scattering situations, Eq. (11.22) reduces to

$$g \approx 1 - \frac{1}{2} \theta_{rms}^2 \tag{11.23}$$

where θ_{rms} is the rms scatter angle of the scalar phase function $P(\theta, \omega)$ given by the equation:

$$\theta_{rms} = \frac{1}{4\pi} \int_0^{2\pi} \int_0^\pi \theta^2 P(\theta, \phi) \cos\theta \, d\Omega \tag{11.24}$$

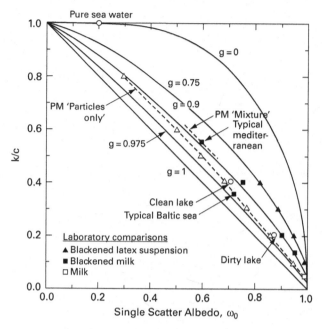

Figure 11.5 Plot of k/c as a function of single scatter albedo for various asymmetry factors [2].

As the value of g approaches 1, the more forward-scattering the medium; conversely, the closer g is to -1, the more backward-scattering the medium. Isotropic scattering media possess asymmetry factors near zero. The graph illustrates the relative magnitude of diffusive scattering for various media. For real ocean waters, the ratio of k to c is typically between 0.4 and 1 for asymmetry factors > 0.75.

As we saw in Chapter 5, Jerlov devised a classification scheme for oceanic waters that divides them into five specific classes: Type I (cleanest), IA, IB, II, and III (dirtiest) [2]. Figure 5.11 and Figure 5.12 show the diffuse attenuation coefficient and water reflectivity for the Jerlov water types, respectively. To benchmark the reader in terms of environment they may know, Hawaii and Caribbean waters are typically Jerlov IB, San Diego water is Jerlov II and Seattle water is Jerlov III. Figure 11.6 and Figure 11.7 depict updated graphs of Jerlov water types and their worldwide distribution, respectively. The Office of Naval Research has created The World-wide Ocean Optics Database (WOOD) for inherent, apparent, and other properties of the various oceans in the world, which can be accessed at http://wood.jhuapl.edu/wood/.

Alternatively, Lee *et al.* have investigated optical propagation in the diffusive regime of seawater using Monte Carlo techniques [5], and confirmed that the upwelling and downwelling radiances profiles are symmetric, approximating a Gaussian distribution for both, and decaying at the same rate both ways. Specifically, they found that

$$N(\theta) = \frac{1}{\pi \theta_{rms}^2} \exp \left[-\left(\theta^2 \Big/ \theta_{rms}^2 \right) \right] \tag{11.25}$$

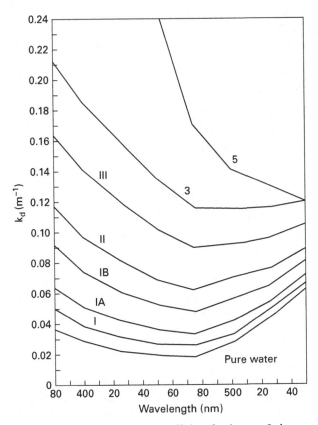

Figure 11.6 Diffused attenuation coefficient for the new Jerlov water types as a function of wavelength [2].

where $\theta_{rms} = 0.6539$ radians (37.8°). To check the validity of this expression, a comparison of the normalized received power for various field-of-view of an underwater receiver at depth and field measurements has been done [3]. The normalized received power is given by

$$f_{sg}\left(\theta_{fov}^{w}\right) = \frac{2\pi \int_{0}^{\theta_{fov}^{w}} N(\theta)\, d\cos\theta}{2\pi \int_{0}^{\pi/2} N(\theta)\, d\cos\theta} \tag{11.26}$$

$$= 1.077311\, I_0 - 0.178888\, I_1 + 0.008198\, I_2 \tag{11.27}$$

with

$$I_0 = 1 - \exp\left(-\theta_{fov}^{w\,2}\big/0.4735336\right) \tag{11.28}$$

$$I_1 = \left(\theta_{fov}^{w\,2}\right)[1 - I_0] + 0.435336\, I_0 \tag{11.29}$$

$$I_2 = \left(\theta_{fov}^{w\,4}\right)[1 - I_0] + 0.435204\, I_1 \tag{11.30}$$

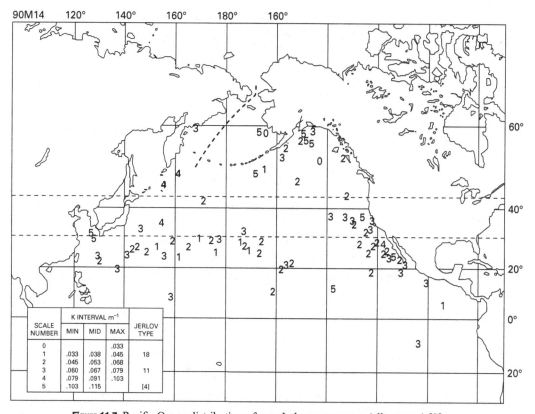

Figure 11.7 Pacific Ocean distribution of new Jerlov water types (all seasons) [2].

and θ_{fov}^{w} being the receiver field-of-view in water. Figure 11.8 compares Eq. (11.27) with experimental data taken from reference [14]. It is apparent from this figure that the comparison between model and data is quite good. Let us now turn to propagation through the air/sea interface and the effect of angle of incidence on diffuse propagation.

Again from the above Monte Carlo work, Lee *et al.* developed an "empirical" relationship for the effect of angle of incidence [5] on received radiance at depth. Specifically, they found that S_1 above is proportional to $F_w(\theta_{inc})$, where

$$F_w(\theta_{inc}) = 1 - (9.72 \times 10^{-4})\,\theta_{inc} - (4.28 \times 10^{-4})\,\theta_{inc}^{2}$$

$$+ (6.04 \times 10^{-6})\,\theta_{inc}^{3} - (4.28 \times 10^{-8})\,\theta_{inc}^{4} \tag{11.31}$$

Figure 11.9 compares this equation with irradiance taken from experimental lake data [15]. Again, we see a very good fit.

Gordon was able to develop an "empirical" equation for the transmittance of incident light through the air/sea interface, which was applicable to a large range of surface wind speeds [16]. Specifically, the air/sea interface transmittance is given by

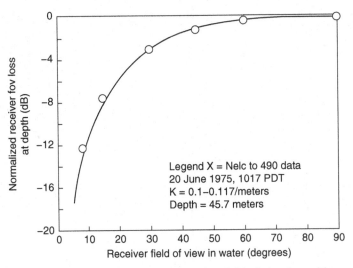

Figure 11.8 Comparison of normalized receiver field-of-view loss with experimental data [2].

$$L_{a/s}(\theta_{inc}) = (0.97186) - (0.11761 \times 10^{-4})\,\theta_{inc} - (0.65066 \times 10^{-4})\,\theta_{inc}^{\,2}$$

$$- (0.2111011 \times 10^{-5})\,\theta_{inc}^{\,3} - (0.19292 \times 10^{-7})\,\theta_{inc}^{\,4} \tag{11.32}$$

with θ_{inc} being angle of incidence as before. This factor also is proportional to S_1.

For cloudy day illumination of the ocean surface, S_1 is proportional to the following effective transmittance term, $L_{a/s}^*(\theta_{inc})$:

$$L_{a/s}^*(\theta_{inc}) = \int_0^{\pi/2} T_{a/s}(\theta)\,F_w(\theta_w)\cos\theta\,d\theta \tag{11.33}$$

where

$$\theta_w = \sin^{-1}(\theta/1.33) \tag{11.34}$$

11.3.3   Better estimation of the water column radiance in the asymptotic scattering case

In Chapter 5, a solution to the radiative transport equation was introduced for small-angle scattering as

$$N_{AH}(\theta_r, \theta_\phi, r) \approx \frac{\exp\left[-a\,z - \left(\dfrac{(\theta_r - \theta_m)^2 + \theta_\phi^{\,2}}{U_\phi^{\,2}}\right) - \dfrac{r^2}{R_1^{\,2}}\right]}{\left(\pi\,U_\phi R_1\right)^2} \tag{11.35}$$

where

$$U_\phi^{\,2} = b z^3\,\theta_{rms}^{\,2}/2 \tag{11.36}$$

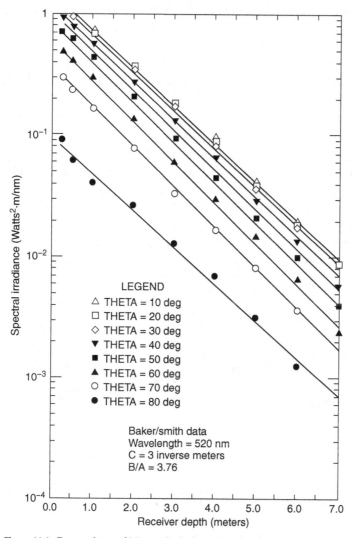

Figure 11.9 Comparison of Monte Carlo-based radiance model estimates with experimental data [11.2].

$$R_1^2 = 2bz^3\theta_{rms}^2/3 \tag{11.37}$$

$$\theta_m = 3r/2z \tag{11.38}$$

[17–19]. Stotts showed that the radiative transfer equation developments associated with small-angle scattering created the radiance distributions for the diffusion regime rather than the non-diffusive scattering [20]. This is because the large scattering angles of most real-world particle distribution (10s of degrees) violated the small-angle scattering (Born approximation).

Integrating Eq. (11.34), Karp was able to estimate the irradiance at depth; namely, he found that

$$\frac{I(0, 0, z)}{I(x_0, y_0)} \approx 2 \int_0^\infty \frac{\exp\left[-a\sqrt{z^2 + r_0^2} - \left(\dfrac{r_0^2}{(2/3\, b\,\theta_{rms}^2)\,\sqrt[3]{z^2 + r_0^2}}\right)\right]}{(2/3\, b\,\theta_{rms}^2)\,\sqrt[3]{z^2 + r_0^2}}\, r_0\, dr_0$$

(11.39)

[18]. Figure 11.10 shows a comparison between Eq. (11.39) and $\exp[-kz]$ (diffusion) for typical blue and green wavelengths. The inherent and apparent properties were defined by assuming the Mediterranean and Baltic Sea numbers in Figure 11.5 to represent the Jerlov IB and II classes, respectively. This figure shows that these two models compare within a few of decibels (dB), good enough for engineering analyses.

Let us now develop an analytical relationship for k/c. Using the numbers in Table 11.1, we derive the following equation:

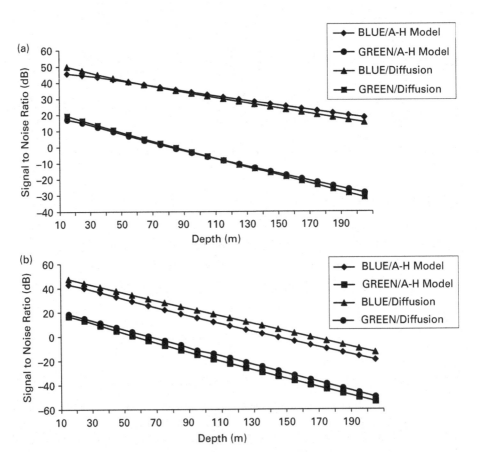

Figure 11.10 Comparison of Arnush-Heggestad and diffusion models for (a) Jerlov IB and (b) Jerlov II water types.

Table 11.1 Inherent and apparent properties of Jerlov IB and II waters for blue (455 nm) and green (532 nm) wavelengths

	Jerlov IB	Jerlov II
Blue laser propagation		
Diffuse attenuation coefficient =	0.034	0.07
Volume extinction coefficient =	0.06	0.2
Volume scattering coefficient =	0.037	0.14
Volume absorption coefficient =	0.023	0.06
RMS scatter angle =	0.5	0.316
Average cosine	0.875	0.95
Green laser propagation		
Diffuse attenuation coefficient =	0.056	0.08
Volume extinction coefficient =	0.1018	0.08
Volume scattering coefficient =	0.0611	0.16
Volume absorption coefficient =	0.0407	0.0686
RMS scatter angle =	0.5	0.316
Average cosine	0.875	0.95

$$\frac{k}{c} \approx 7\,(1-\omega_0)\,(1-g) \tag{11.40}$$

This may be useful in relating the inherent and apparent optical properties for general link budget analysis using the Jerlov water types.

11.4 System design models for the optical scatter channel

Multipath pulse spreading by clouds creates the most stressing laser communications situation as the laser energy is spread over a few to 10s of microseconds, allowing solar radiation or other external optical sources to enter an optical receiver and interfere with the detection process. One of the best ways to minimize background radiation like solar illumination corrupting communications system performance is peak power detection, which only measures all incoming optical energy within approximately the pulsed laser signal's time duration, its pulse-width. The most natural way to exploit this energy detection approach for communications is through pulse position modulation (PPM). We discussed this in Chapter 6 when we introduced PPM.

PPM conveys information by varying the time of occurrence of a single optical pulse within a data sample time period. The sample period, called a frame, is subdivided into several time windows, called slots. Figure 11.11 shows this approach, with the key PPM parameters defined; specifically for octary PPM systems using a pulsed laser. Shown in this figure is the receipt of a pulse in two adjacent frames. The pulse in the first frame and the one in the second frame would translate into the binary words (011) and (110), respectively.

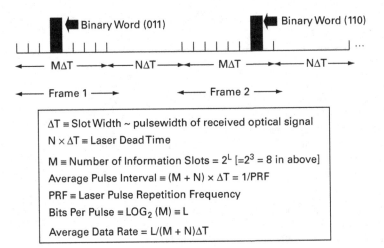

Figure 11.11 Diagram of pulse position modulation and key parameter definitions.

The figure contains the following annotations:

Binary Word (011)   Binary Word (110)

$M\Delta T$ $N\Delta T$ $M\Delta T$ $N\Delta T$

Frame 1   Frame 2

$\Delta T \equiv$ Slot Width ~ pulsewidth of received optical signal

$N \times \Delta T \equiv$ Laser Dead Time

$M \equiv$ Number of Information Slots $= 2^L$ [$=2^3 = 8$ in above]

Average Pulse Interval $\equiv (M + N) \times \Delta T = 1/PRF$

$PRF \equiv$ Laser Pulse Repetition Frequency

Bits Per Pulse $\equiv LOG_2 (M) \equiv L$

Average Data Rate $= L/(M + N)\Delta T$

11.3.1   Maximizing system data rate

To maximize the data rate requires us to maximize the equation

$$\frac{\log_2 M}{(M + N)\,\Delta T} \tag{11.41}$$

To do this, we set

$$T_D = N\,\Delta T \tag{11.42}$$

$$\Delta T = \alpha\,T_D \tag{11.43}$$

where we have defined the communications slot width in terms of the laser dead time. Here, α is a system constraint that is imposed upon the design. The pulse repetition can be written as

$$R_p = \frac{1}{T_D\,(1 + \alpha M)} \tag{11.44}$$

for

$$\alpha = \frac{\Delta T}{T_D} \tag{11.45}$$

and

$$R_{p_{\max}} = \frac{1}{T_D} \tag{11.46}$$

for $\alpha = 0$. This gives an average data rate equal to

$$R = R_p \log_2 M \tag{11.47}$$

$$= R_{p_{\max}} \frac{\log_2 M}{(1 + \alpha M)} \tag{11.48}$$

or in terms of the maximum pulse rate

$$\frac{R}{R_{p_{max}}} = \frac{\log_2 M}{(1 + \alpha M)} \equiv \beta < \log_2 M \qquad (11.49)$$

Differentiating Eq. (11.49) with respect to the number of bits per pulse and setting the result to zero yields

$$\alpha = \frac{\log_2 e}{M \log_2 (M/e)} \qquad (11.50)$$

Figure 11.12 shows the ratio of slot width to dead time as a function of the optimum word length in bits per pulse. Figure 11.13 gives the maximum normalized data rate β_{max}, as a function of the optimum word length in bits per pulse. It is apparent from this last figure that the maximum normalized data rate has an asymptotic trend of $\log_2 (M/e) < \log_2 M$

Example 11.1 Let us calculate the maximum data rate. Substituting Eq. (11.47) into Eq. (11.48), we have

$$\frac{R}{R_{p_{max}}} = \frac{\log_2(M)}{(1 + \alpha M)} = \frac{\log_2(M)}{\left(1 + \left[\dfrac{\log_2(e)}{M \log_2(M/e)}\right] M\right)}$$

$$= \frac{\log_2(M)[M \log_2(M/e)]}{(M \log_2 (M/e) + M \log_2(e))} = \frac{M \log_2(M)\log_2(M/e)}{M \log_2(M)}$$

$$= \log_2(M/e) \qquad (11.51)$$

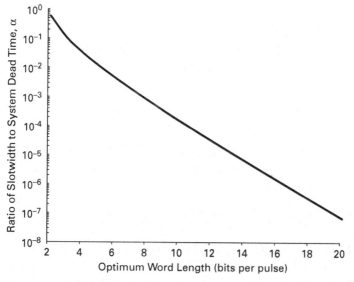

Figure 11.12 Ratio of Slotwidth to System Dead Time as a Function of the Optimum Word Length (bits per pulse).

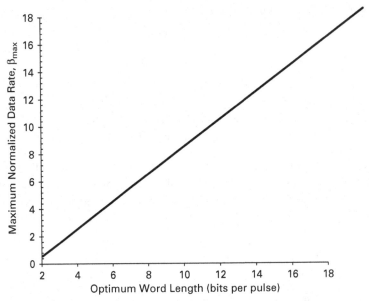

Figure 11.13 Maximum normalized data rate, β_{max}, as a function of the optimum word length (bits per pulse).

Under solar background limited operations, the high background noise makes the noise process tend towards a Gaussian random process. The probability of making an error in demodulating a PPM symbol is given by

$$P_E^{symbol} = 1 - \frac{1}{\sqrt{2\pi}} \int_{-\infty}^{\infty} \exp\left[-(x - \text{SNR})^2/2\right] (1 - \text{erfc}[x]) M^{-1} dx, \qquad (11.52)$$

which can be approximated by

$$P_E^{symbol} \approx \frac{M-1}{\sqrt{\pi \, \text{SNR}}} \exp\left[-\text{SNR}/4\right] \qquad (11.53)$$

where

$$\text{SNR} = \frac{K_s^2}{K_N^2} \qquad (11.54)$$

Figure 11.14 illustrates the probability of symbol error as a function of signal-to-noise ratio. Since we are interested in the probability of bit error, Eq. (11.22) converts to this probability using the relation

$$P_E^{bit} \approx \frac{2^{\ell-1}}{2^{\ell}-1} P_E^{symbol} \qquad (11.55)$$

Figure 11.15 gives the probability of bit error as a function of signal-to-noise ratio per bit after recomputing the previous result. Notice that as we go to larger values of $M = 2^{\ell}$, the signal-to-noise ratio per bit decreases for the same performance. For the Gaussian

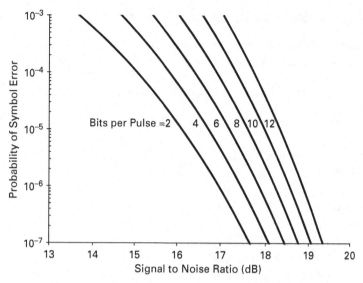

Figure 11.14 Probability of symbol error as a function of signal to noise ratio.

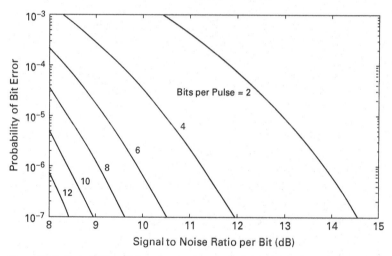

Figure 11.15 Probability of bit error as a function of signal to noise ratio per bit.

channel, this ratio approaches -1.59 dB, the Shannon limit, as $M \to \infty$. For the smaller values of M, one also can get a reduced per bit signal-to-noise ratio by using FEC capability as we suggested in the turbulence channel chapter. As we found previously, these codes can yield several decibels of improvement with modest complexity. However, one needs at least a 6 to 6.5 dB per bit signal-to-noise ratio in order to obtain a 3 dB improvement if, say, $l = 8$ is selected.

11.3.2   Signal-to-noise ratio in the optical scatter channel

The general signal-to-noise ratio equation for the optical scatter channel where the laser transmitter is in the atmosphere and the receiver is submerged in water is equal to

$$\text{SNR}_{Electrical}^{OpticalScatterChannel} = \text{SNR}_E^{OSC} = \frac{(\eta/hv)\,(P_{Tx}/\text{PRF})^2}{P_{Tx}/\text{PRF} + P_{Noise}\,\Delta T} \tag{11.56}$$

where

$\eta \equiv$ detector quantum efficiency
$h \equiv$ Plank's constant $= 6.626 \times 10^{-34}$ joule-seconds
$v \equiv$ frequency of optical radiation $= c_0/\lambda$
$c_0 \equiv$ speed of light $\approx 3.0 \times 10^8$ meters per second
$\lambda \equiv$ wavelength of optical radiation
$P_{Tx} \equiv$ received laser power
$\text{PRF} \equiv$ pulse repetition frequency
$P_{Noise} \equiv$ received background noise power
$\Delta T \equiv$ pulse width of received optical signal.

From previous work [1,2,21], we know that

$$P_{Tx} \approx \gamma_{Tx}\,\gamma_{Rx}\left(\frac{d_R^{\,2}}{d_S^{\,2}}\right)L_a\,L_c\,L_{a/s}\,L_w f_{sg}\left(\theta_{fov}^w\right)P_{ave} \tag{11.57}$$

and

$$P_{Noise} \approx \gamma_{Tx}\,\gamma_{Rx}\,L_a\,L_c\,L_{a/s}\,L_w f_{sg}\left(\theta_{fov}^w\right)I_{solar}\,d_R^{\,2}\,\Delta\lambda \tag{11.58}$$

where

$\gamma_{Tx} \equiv$ transmittance of laser optics
$\gamma_{Rx} \equiv$ transmittance of receiver optics
$d_R \equiv$ diameter of receiver optics
$d_s \equiv$ diameter of incoming laser beam at the receiver aperture
$L_a \equiv$ atmospheric transmittance
$L_c \equiv$ cloud transmittance
$L_{a/s} \equiv$ transmittance through the air/sea interface
$L_w \equiv$ transmittance through water column to depth $D =$

$$F_w\left(\theta_{inc}\right)\exp[-k\,D] \tag{11.59}$$

$F_w\left(\theta_{inc}\right) \equiv$ irradiance correction for off-axis light illumination
$k \equiv$ diffuse attenuation coefficient
$f_{sg}\left(\theta_{fov}^w\right) \equiv$ power received at depth by a finite FOV receiver
$P_{ave} \equiv$ average laser power
$I_{solar} \equiv$ exo-atmospheric solar irradiance
$\Delta\lambda\gamma \equiv$ spectral passband of optical receiver

The form of Eq. (11.56) assumes that all system noise sources are much less than the background and signal-shot noise levels, which is generally true for most applications. That is, the latter two noise sources dominate the signal-to-noise ratio calculation in general. Typically, the solar irradiance at mid-day is 1500 W/(m²-μm) at 532 nm and 1300 W/(m²-μm) at 455 nm. At nighttime, we would replace I_{solar} with I_{lunar} where

$$I_{lunar} = 10^{-6} I_{solar} \qquad (11.60)$$

All other communications links geometries will be a variant of the above, which the readers can easily modify to fit their applications.

Example 11.2 Let us look at a link budget for a potential system design. We will assume a 100 watt Nd:YAG laser-based system operating at a PRF of 2500 Hz. Table 11.2 shows a set of example Green optical receiver and filter parameters used in previous system demonstrations [21–23].

For this example, we will assume both the transmitter, Sun and Moon are always overhead and the atmospheric loss is

$$L_a = \exp[-0.357] \qquad (11.61)$$

Figure 11.16 shows a typical cumulative distribution for cloud optical thickness based on meteorological measurements [2 , p. 361 and Appendix E]. In this case, the distribution is for the area east of Hawaii. Typically, the physical thickness of a cloud layer is 25 times the cloud optical thickness in meters, e.g., for a $\tau = 30$ cloud, the physical thickness of the layer is 750 meters. Referring to this figure, we see that clear sky conditions occur approximately 65% of the time and clear sky / thin cloud (i.e., $\tau \leq 30$) conditions around 72% of the time. For high link availability, one needs to be able to overcome thick cloud conditions above 100. For example, a 90% link availability implies communications through clouds of optical thickness, τ, of 100 or less; for 95% link availability, one must be able to communicate through $\tau = 150$ clouds.

Figure 11.17 and Figure 11.18 show the received signal-to-noise ratio as function of receiver depth for daytime/nighttime operation in Jerlov IB and Jerlov II waters under no cloud, clouds with $\tau = 100$ and clouds with $\tau = 150$. We will assume a nominal 13 dB communications threshold (assumes FEC coding). The results shown are not too

Table 11.2 Example optical receiver and filter parameters for green (532 nm) wavelengths

Parameters / filter technology	Quartz green	Green dichroic
Receiver clear aperture (m²)	0.5654	0.5654
Transmittance	10%	10%
Spectral bandwidth (angstroms)	3	5
Quantum efficiency	0.3	0.3
FOV (degrees in air)	± 45	± 45
Temporal response	N/A	N/A

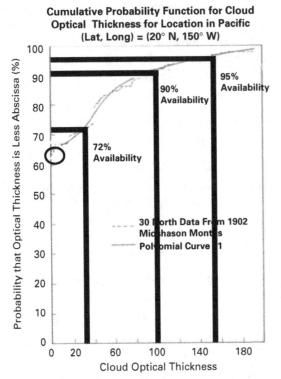

Figure 11.16 Probability of cloud optical thickness being less than abscissa off the Eastern Coast of Hawaii.

surprising. Figure 11.17(a) shows that the link closes to significant depths with no clouds, but does not close under the $\tau = 100$ or 150 conditions. This is because multipath time spreading causes the nominal 40 ns pulse of the original laser pulse to grow to 10s of microseconds (or larger). This results in the system peak power at the receiver going down. To counter this, one needs better filter technology that can spectrally narrow the optical bandwidth of the optical filters, but still keep a wide angular field-of-view, e.g., atomic line filters [23], given the larger pulse width under cloudy conditions. On the other hand, the system works much better when the background light at the receiver is reduced. This can be seen in Figure 11.17(b) where the system is quantum-limited, even under full moonlight illumination. This plot shows the system can reach appreciable depths in open ocean water (Jerlov IB) at night. Figure 11.18 shows similar performance, but the receiver depths for no clouds drop 100 meters during the day and 50 meters during the night because of the dirtier littoral water (Jerlov IB) the light must traverse.

Example 11.3 For clear sky conditions, Table 11.3 shows the data rates at the reported depths for a 75 watt 62.6 Hz green laser system operating in the daytime under Jerlov IB water conditions. The spot at the cloud top was 5 km. The communications threshold was set to 22–25 dB. Two things are clear from this table. First, using a lower PRF laser creates

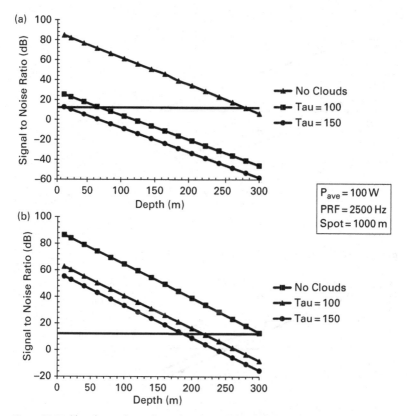

Figure 11.17 Signal to noise ratio as function of receiver depth for (a) daytime and (b) nighttime operation in Jerlov IB water.

better depth penetration over what we saw in the Figure 11.17, even at the higher communications threshold. The reason is that the received peak power is much larger, allowing the system to overcome the solar background noise better. Second, the multipath time spreading for this laser subsystem limits the data rate to around a few kilobits per second. Alternate designs could yield higher data rates like a megabit per second because the multipath time spreading in the water channel for depths up to 300 m is around 200 ns, or less. However, in this situation, for those applications where connectivity at appreciable depth is paramount, the capabilities to do "chat", texting, canned messages and other low-data-rate communications are the only ones to be expected.

Table 11.4 shows the resulting data rates at the reported depths for a 75 watt 62.6 Hz green laser system operating in the daytime under open ocean and cloudy conditions. The spot at the cloud top again was 5 km. We see that a lower PRF laser also causes better depth penetration over what we saw in the Figure 11.17, similar to what we also saw above. However, the multipath time spreading limits the system data rate to around half a kilobit per second when clouds get in the way. This may be good enough for chat, texting, canned messages and other low-data-rate communication, but no alternate design is going to change things.

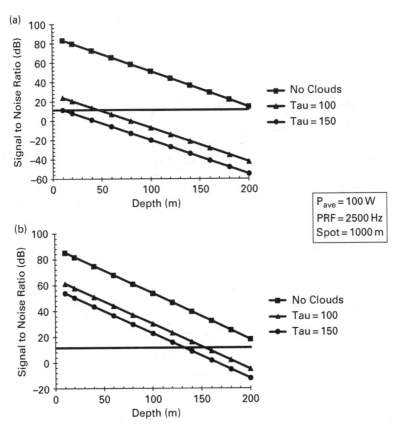

Figure 11.18 Signal to noise ratio as function of receiver depth for (a) daytime and (b) nighttime operation in Jerlov II water.

Table 11.3 Data rates for a 75 watt 62.5 Hz green laser communications system operating in the optical scatter channel under clear sky conditions

PRF	Depth	Data rate	Bits/pulse
62.5 Hz	~250	1.00 Kb/s	16
125 Hz	~230	1.88 Kb/s	15
250 Hz	~210	3.75 Kb/s	15

Table 11.4 Data rates for a 75 watt 62.5 Hz green laser communications system operating in the optical scatter channel for three cloud thickness conditions

Cloud	Depth	Data rate	Bits/pulse
Tau = 30	~170 m	688 b/s	11
Tau = 100	~100 m	563 b/s	9
Tau = 150	~60 m	500 b/s	8

Example 11.4 Let us see how an FEC capability can help the design of a laser communication system operating between a submerged research vehicle and a loitering aircraft. We will assume that they can exchange chat information or data using one common wavelength, and both transmitters use atomic clocks independently to synchronize. The uplink laser operates at 125 pulses per second with a frame time of 8 ms. We further assume the downlink laser operates at 500 pulses per second and in order to avoid receiver damage must use 500 blanking pulses with a 140 µs footprint in any 8 ms interval. Consequently, there will be four collisions in each frame. This creates a collision rate with the received information of $(1.4 \times 10^{-4}/8 \times 10^{-3}) \approx 1/57$, which is essentially the probability of overlapping a signal pulse. Since the chance of a blanking pulse interfering with an uplink frame is one in 57, we must accept the fact that one in 57 symbols will be lost. To maximize the data rate PPM will use under a symbol error rate of $P_E = 1/57$, we need to correct a minimum number of errors to maintain a high data rate.

With regard to chat, the average typing speed of a professional typist is about 50–70 words per minute, and half of that for a "hunt and peck" typist. If we get into speech translation, spoken English can be up to 300 words per minute, with an average word length of five characters. Allowing each character to have $26 + 10$ variants, we need at least a 6-bit character. To cover all bets, we should be able to handle a minimum character rate from 5 to 25 per second. Since our rate is 125 pps, and we must correct symbols, Reed-Solomon codes (RS) with $n = 127, 63, 31, 15$ characters would suffice. We also wish to maximize the data rate and minimize the signal power. The probability of symbol error for an (n, k) RS code, t errors corrected and $k = n - 2t$ information symbols, is given by

$$PE_t = \frac{\sum_{j=t+1}^{n} j \frac{n! \cdot p^i (1-p)^{n-j}}{j! \cdot (n-j)!}}{n} \tag{11.62}$$

$$= \frac{\sum_{j=t+1}^{2^m-1} j \frac{(2^m-1)! p^j (1-p)^{2^m-1-j}}{j!(2^m-1-j)!}}{2^m-1} \tag{11.63}$$

where $n = 2^m - 1$, and $k = 2^m - 1 - 2t$ [24].[1] To correct $t = 1$ and $t = 2$ errors, the (127, 125), (127, 123), (63, 61), (63, 59), (31, 29), (31, 27), and (15, 13), (15, 11) RS codes are good candidates as a starting point, delivering 125, 123, 61, 59, 29, 27, and 13, 11 symbols, respectively, at symbol rates of

---

[1] For air-to-subsurface communications under cloudy conditions, two applications jump out, data and chat. For the data, it might be interesting to consider an inner code, where each symbol contains an additional degree of coding. In this example, we will only consider the outer RS code and its application to chatting.

$$125\left(125/127\right)=123\,,\quad 123\left(125/127\right)=121$$

$$61\left(125/63\right)=121\,,\quad 59\left(125/63\right)=117$$

$$29\left(125/31\right)=117\,,\quad 27\left(125/31\right)=109$$

$$13\left(125/15\right)=108\,,\quad 11\ \left(125/15\right)=92$$

symbols per second, respectively. Figure 11.19 depicts the output symbol error as a function of input symbol error for the above RS codes. Using an input $P_E^{Input\ Symbol}=1/57$ (horizontal line), the output symbol errors for each of the above RS codes also are listed in Figure 11.19. It is clear the longer block codes do not provide much improvement, but things get better as the codes shorten and a double error correction is used.

When a block error is made, k symbols are lost, and this happens $k\left(125/n\right)P_E$ times per second, so that the time between symbol errors τ becomes

$$\tau=\frac{1}{k{\cdot}\left(125/n\right){\cdot}P_E}\tag{11.64}$$

$$\frac{1}{\left(\dfrac{\text{symbols}}{\text{block}}\right)\left(\dfrac{\text{blocks}}{\text{second}}\right)\left(\dfrac{\text{errors}}{\text{symbol}}\right)}=\frac{1}{\left(\text{errors/sec}\right)}=\text{sec/error}\tag{11.65}$$

Using output symbol error rates from Figure 11.19, the times between errors for the above RS codes are 0.51, 0.69, 0.69, 1.53, 1.12, 4.83, 2.25 and 23.1 seconds, respectively. As with the output symbol error improvement, shorter codes and double error correction improve performance; here extending the time between errors to seconds rather than parts of a second.

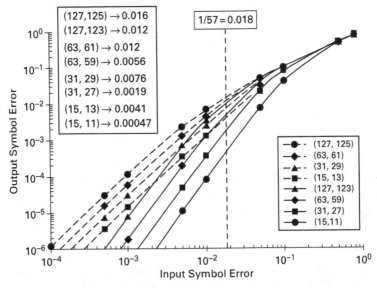

Figure 11.19 Graph of output symbol error as a function of input symbol error for various Reed Solomon (n, k) codes.

At this point, it is necessary to determine what modes of operations are necessary. For example, if correctness is important and we want the average time between errors to be an hour (3600 seconds), we see that we need $P_E(k^{125}/n) \approx 2.77 \times 10^{-4}$ from Eq. (11.64). This means that it must be range from 1.98×10^{-6} to 2.7×10^{-6} for the above RS codes to meet this goal. We now need to look at the probability of error in this case; for simplicity, we will assume the lower level, $\sim 2 \times 10^{-6}$. The probability of t errors is approximately given by the following equation:

$$P_t \approx \left(\frac{1}{57}\right)^{t+1} + \sum_{k=1}^{t+1} PE_k \left(1 - \frac{1}{57}\right)^{k} \left(\frac{1}{57}\right)^{t+1-k} \tag{11.66}$$

Figure 11.20 illustrates this probability as a function of input symbol error for the RS codes (63, 61), (63, 59) and (63, 57), examples chosen for illustrative purposes. These three RS (63, k) codes represent single, double and triple error detection codes. Assuming the probability of t errors equals the above 2×10^{-6}, it is clear from this figure that single and double error detection cannot meet that number. For an average time between symbol errors of one hour, we must employ triple error detection. This means that a message is received incorrectly when four errors occur. Referring to Figure 11.20 again, we see that the 2×10^{-6} level is achieved by an RS (63,57) code when the input symbol error equals 0.0054, a factor of 3 less than the value of 1/57 previously used, which gives a probability of t errors of $\sim 3.6 \times 10^{-4}$. The penalty for this decrease in input symbol error is that the PPM signal-to-noise ratio must increase approximately 2 dB, which can be seen in Figure 11.21, assuming the probability of error curve for $M = 64$ is

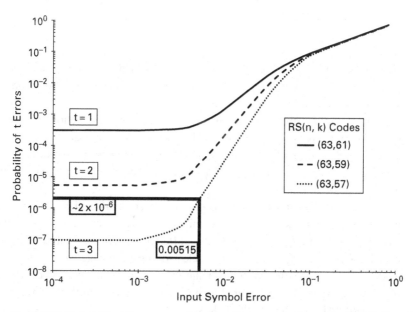

Figure 11.20 Probability of t-errors as a function of input symbol error for a RS(63, k) codes to detect 1, 2 and 3 errors.

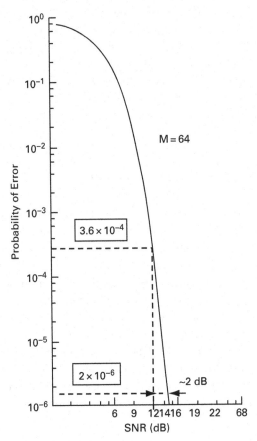

Figure 11.21 Probability of errors as a function of signal-to-noise ratio (SNR) for $M=64$.

essentially the same curve as for $M = 63$ [25]. This is a minor change in system performance as compared to the decrease in the time between symbol errors from 3600 seconds to around 25 seconds. Given the above, we need only to feed back a short message to the transmitter to retransmit to facilitate error correction. This requires four symbol errors to pass through undetected, which takes an average of one hour. The effect is to add retransmit symbols to the data stream, which will cost approximately 0.5 second ($\sim$63/125) every time we retransmit. This again is only a minor performance impact in the greater scheme of things.

We also might point out that the majority of system errors will come from jitter and the use of a Grey code as the inner code might increase the errors corrected by one. This would work by making the PPM slots different in only one bit location from each adjacent slot. Then when the error is detected, testing for the word in the slot on both adjacent sides would test positive if the error arose from jitter. If computation time permits, testing of two adjacent slots on each side would further help if the blanking pulse straddled two adjacent PPM slots.

11.5   Over the horizon optical communications

Besides direct line-of-sight (LOS) optical scatter links, one also can use this mechanism to achieve optical communications over the horizon (OTH). Optical communications below LOS (OTH) has been a research topic for the past few decades, but serious implementation has yet to occur. In this section, we will discuss this type of communications, providing a first-order model for characterizing its performance. In addition, we will compare this model with experimental data.

OTH optical propagation can occur using a number of different physical mechanisms [2, pp. 289–293]. For example, light can be redirected to a below-LOS receiver using optical ducting, or simple downward curvature of the initial beam. Radiation transfer by atmospheric aerosols and transfer by cloud bottoms are two other means by which one can achieve below-LOS link connectivity. The radiative transfer equation in the small-angle scattering approximation has been used by Levine and O'Brien to describe atmospheric aerosol-induced optical propagation for below-LOS communications [26], the results of which will be summarized here.

Figure 11.22 depicts the OTH propagation mechanisms proposed by Levine and O'Brien. They distinguish two basic mechanisms. One is particulate single scatter and the other is particulate multiple scatter. In general, one would expect the latter mechanism to dominate the former at ranges in excess of a few scattering lengths due to the lack of absorption in the atmosphere at most visible and NVIR wavelengths [2, pp. 221–237]. However, the proximity of the link to the Earth's surface affects optical propagation, with path loss similar to that of absorption, i.e., light is lost forever. Given those facts, Levine and O'Brien proposed two propagation models for OTH communications; one based on single scattering of radiation and the other on forward multiple scattering. In the former case, they showed that the integrated path loss (ratio of received power to transmitter power) is given by

$$\frac{P_R}{P_T} = \frac{8Ac\exp(-cD)}{3\pi\lambda r} \sum_{i=1}^{3} \frac{c_i}{b_i^3} \left(J(x_i^{max}) - J(x_i^{min}) \right) \bigg/ \sum_{j=1}^{3} \frac{c_j}{b_j^2} U(b_j) \qquad (11.67)$$

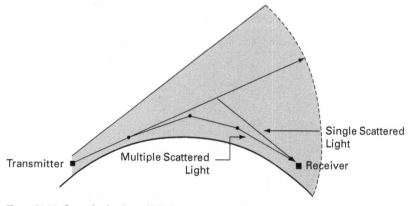

Figure 11.22 Over-the-horizon (OTH) propagation phenomena (cloud-free).

where

$$x_i^{\max} = \frac{2\pi}{\lambda b_i}(\theta_{\min} + 2\theta_{FOV})x_i^{\min} = \frac{2\pi}{\lambda b_i}\sin(\theta_{\min}) \tag{11.68}$$

$$J(x) = L(x) - \frac{2}{(1+y)^3}L\left(\frac{x}{1+y}\right) + \frac{1}{(1+2y)^3}L\left(\frac{x}{1+2y}\right) \tag{11.69}$$

$$y_i = 2\frac{k}{b_i}\left|\frac{m^2-1}{m^2+2}\right| \tag{11.70}$$

$$L(x) = \left[\left(1 - \frac{5}{2}W + \frac{5}{2}W^2\right)K(W) - (1-2W)E(W)\right]W^{-3/2} \tag{11.71}$$

$$U(b) = 1 - \frac{4}{\left(1+\frac{g}{b}\right)^2} + \frac{6}{\left(1+\frac{2g}{b}\right)^2} - \frac{4}{\left(1+\frac{3g}{b}\right)^2} + \frac{1}{\left(1+\frac{4g}{b}\right)^2} \tag{11.72}$$

$$g = k\sqrt[4]{4/3}\sqrt{\left|\frac{m^2-1}{m^2+2}\right|} \tag{11.73}$$

$$b_i = \frac{d_i}{F'} \tag{11.74}$$

$$K(x) = \int_0^{\pi/2}\frac{d\theta}{\sqrt{(1-x\sin^2(\theta))}} \tag{11.75}$$

$$E(x) = \int_0^{\pi/2}d\theta\sqrt{(1-x\sin^2(\theta))} \tag{11.76}$$

$$W = 4x^2/(1+4x^2) \tag{11.77}$$

$$k = \frac{2\pi}{\lambda} \tag{11.78}$$

$$F' = F(RH)/F(80) \tag{11.79}$$

$$F(RH) = 1 - 0.9\ln(1 - RH/100) \tag{11.80}$$

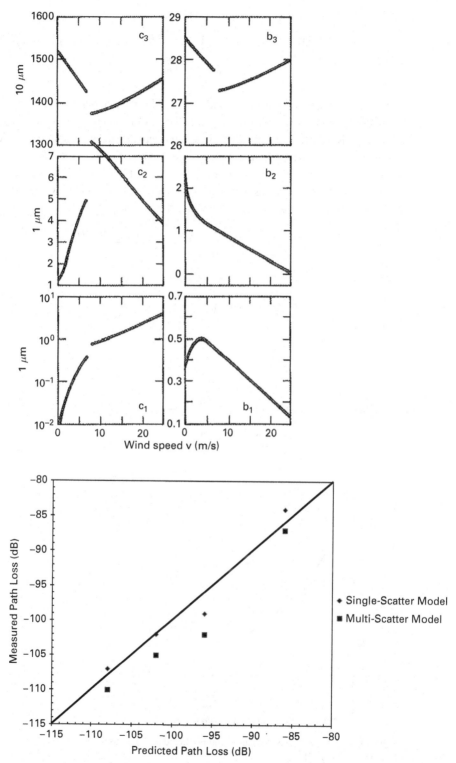

Figure 11.23 Coefficients, (c_i, b_i; i=1, 2, 3) for OTH single-scatter model [25]. Comparison of measured and predicted path loss for 63 km OTH range [25].

$$r = \left(D^2/2R\right)\left[\left(1-I_T\right)^2 - I_R{}^2\right] + D\alpha\left(1-I_T\right) \tag{11.81}$$

$$I_{T,R} = \sqrt{2\,h_{T,R}\left({}^R/_D\right)} \tag{11.82}$$

In the above equations, D is the ground range, RH is the percent relative humidity, θ_{FOV} is the half-angle field of view of the receiver in the elevation plane, A is the area of the receiver aperture, m is the complex refractive index of the aerosols at the transmitter wavelength, α is the elevation angle of the transmitter, R is the effective radius of the Earth, and $h_{T,R}$ is the height of the (transmitter, receiver) above sea level. Figure 11.23 shows the coefficients $(c_i, b_i; k = 1, 2, 3)$ for various wind speeds and a relative humidity level of 80%.

For the latter case, Levine and O'Brien gave the integrated path loss as

$$\frac{P_R}{P_T} = \frac{2,515\,A\,p(\gamma)p(0)\,c\exp\left(-\tau_0 - \tau_1\right)}{D_1} \tag{11.83}$$

where

$$\gamma = \frac{D}{R} + \sin^{-1}\left(1 + \frac{h_T}{R}\right) + \sin^{-1}\left(1 + \frac{h_R}{R}\right) \tag{11.84}$$

$$\tau_0 = c\,Z_0 \tag{11.85}$$

$$\tau_1 = c\,D_1 \tag{11.86}$$

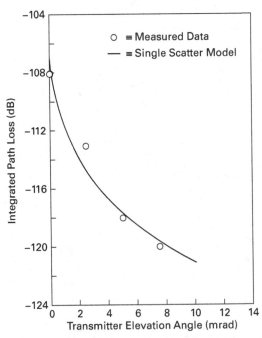

Figure 11.24 Comparison of measured and predicted path loss as a function on elevation angle for 63 km OTH range [25].

$$Z_0 = \sqrt{2 R h_T} + R \tan(\gamma) - R(\sec(\gamma) - 1) \csc(\gamma) \tag{11.87}$$

$$D_1 = \sqrt{2 R h_R} + R(\sec(\gamma) - 1) \cot(\gamma) \tag{11.88}$$

In the above equations, $p(\gamma)$ is the scalar phase function for the atmosphere, γ is the elevation angle.

If both the transmitter and receiver are close to the Earth's surface, then Eq. (11.84) reduces to

$$\frac{P_R}{P_T} = \frac{5.03 \, A \, p(0)^2 \, c \exp(-cD)}{D_1} \tag{11.89}$$

where all the parameters have been defined previously.

Figure 11.23 is a comparison of measured and predicted path loss using both the single-scatter and multi-scatter models for a 63 km link between San Diego, California, and Oceanside, California, taken on 26 February 1976 [27]. This plot suggests the single-scatter predictions compares better with the measured data than the multi-scatter estimates. Figure 11.24 shows the single-scatter model predictions with measured data as a function of transmitter elevation angle. Again, we see reasonable matching between the measured data and predictions. A sensitivity analysis by the authors showed that the estimated path loss could increase (1) by 1.8 dB at all elevations if the volume extinction coefficient increases by 5%; (2) by 0.4–1.4 dB for elevations angles between 0 and 10 mrad if the effective Earth radius decreases by 5%; and (3) by 2.5 dB at all elevation angles if the wind speed estimate decreases from 5 m/s to 1 m/s [26]. In spite of that, they felt the agreement between theory and experiment was still reasonable [27].

References

1. R. S. Kennedy. Communication through optical scattering channels: an introduction. *Proceedings of the IEEE*, 58(10) (1970), pp. 1651–1665.
2. S. Karp, R. M. Gagliardi, S. E. Moran and L. B. Stotts. *Optical Channels: Fiber, Atmosphere, Water and Clouds*, Ch. 5. Plenum, New York (1988).
3. R. W. Preisendorfer. *Hydrologic Optics, Vol. V, Properties*. US Government Printing Office, Boulder, CO (1976).
4. E. A. Bucher. Computer simulation of light pulse propagation for communications through thick clouds. *Applied Optics*, 12 (1973), pp. 2391–2400
5. R G. M. Lee, C. M. Ciany and C. Tranchita, McDonnell-Douglas Astronautics. Private communications.
6. E. Danielson, D. R. Moore and H. C. Van de Hulst. The transfer of visible radiation through clouds. *Journal of the Atmospheric Sciences*, 26 (1969), pp. 1078–1087.
7. G. Tyler and R. W. Preisendorfer. Transmission of energy within the sea. In M. N. Hill (ed.), *The Sea*, Vol. 2. Interscience, New York (1962).
8. H. C. van de Hulst. Multiple Light scattering in atmospheres, oceans, clouds and snow. Institute for Atmospheric Optics and Remote Sensing, Short Course No. 420, Williamsburg, Virginia, December 4–8 (1978).

9. N. G. Jerlov. *Marine Optics*. Elsevier, Amsterdam (1976).

10. A. Morel. Optical properties of pure water and pure sea water. In N. G. Jerlov and E. S. Nielsen (eds), *Optical Aspects of Oceanography*. Academic Press, New York (1974), pp. 1–24.

11. D. B. Judd. Terms, definitions and symbols in reflectometry. *Journal of the Optical Society of America*, 57 (1967), pp. 445–452.

12. R. W. Austin, S. Q. Duntley, W. H. Wilson, C. G. Edgerton and S. E. Moran. In *Ocean Color Analysis*, SIO Reference 74–10 (1974).

13. H. C. van de Hulst. *Light Scattering by Small Particles*. John Wiley and Sons, New York (1957).

14. R. N. Driscoll, J. N. Martin and S. Karp. OPTSATCOM Field Measurements. Technical Document 490, Naval Electronic Laboratory Center (June 1, 1976).

15. K. S. Baker and R. C. Smith. Quasi-inherent characteristics of the diffuse attenuation coefficient for irradiance. *Proc. Photo-Opt. Instrum. Eng.: Ocean Optics VI*, 208 (1980), pp. 60–63.

16. J. Gordon. *Direction Radiance (Luminescence) of the Sea Surface*. SIO Ref. B9–20 (October 1969).

17. D. Arnush. Underwater light-beam propagation in the small-angle scattering approximation. *Journal of the Optical Society of America*, 59 (1969), p. 686.

18. S. Karp. Optical communications between underwater and above surface (satellite) terminals. *IEEE Transactions on Communications*, **24**(1) (1976), pp. 66–81.

19. H. M. Heggestad. Optical communications through multiple scattering media. MIT/RLE Tech. Rpt. 472 (November 1968).

20. L. B. Stotts. Limitations of approximate Fourier techniques in solving radiative transfer problems. *Journal of the Optical Society of America*, **69**(12) (1979), p. 1719.

21. L. B. Stotts. Tactical airborne laser communications. Invited Paper, Proceedings of the IEEE Military Communications Conference 1990, McLean, Virginia, November 1991.

22. L. B. Stotts. Design Considerations of a submarine laser communications system. *Proceeding of the IEEE National Telecommunications Conference '92*, George Washington University.

23. J. J. Puschell, R. J. Giannaris and L. B. Stotts. Autonomous data optical relay experiment. *Proceeding of the IEEE National Telecommunications Conference '92*, George Washington University.

24. G. C. Clark and J. B. Cain. *Error-Correcting coding for Digital Communications*. Plenum Press, New York (1988).

25. A. J. Viterbi. *Principles of Coherent Communications*. McGraw-Hill, New York (1966).

26. P. H. Levine and M. E. O'Brien. ELOS meteorology sensitivity study, Megatek Final Report No. R2005–099-F-1, Contract No. N00123-75-C-0328, Task MEG-TA-009 (November 1977).

27. G. C. Mooradian, M. Geller, P. H. Levine, D. H. Stephens and L. B. Stotts. Over-the-horizon optical propagation in a maritime environment. *Applied Optics*, **19**(1) (1980), p. 11.

Appendix A: Two-dimensional Poisson processes

In this appendix we will derive the Poisson point process over a two-dimensional area which is evolving in time, yielding a three-dimensional volume in x,y,t. This describes the evolving current process from a focal plane array of pixels, given the input intensity, $I_D(x,y;t)$.

Let us again consider Figure 3.1 and the derived statistics

$$P_k(k) = \frac{(m_V)^k}{k!} e^{-m_V} \tag{A.1}$$

where

$$m_V \equiv \alpha \int_V I(v) dv = \alpha \int_A \int_t^{t+T} I\{\rho, \vec{r}\} d\rho d\vec{r} \tag{A.2}$$

Specifically we now derive the joint density of the random sequence $z = (z_1, z_2 \ldots z_k)$ over the volume V, given that there are k random occurrences in that volume. We write this conditional joint density as $p_z(z_1, z_2, \ldots z_k/k)$. This can be derived as follows. Consider the volume V to contain the infinitesimal volumes $(v_1 + \Delta v), (v_2 + \Delta v), \ldots (v_k + \Delta v)$. The probability of getting a Poisson occurrence in each of the infinitesimal volumes above and simultaneously getting exactly k occurrences is simply the probability of getting one occurrence in each infinitesimal volume and not getting any occurrence elsewhere in the volume V. Since the infinitesimal volumes and their external regions are disjoint, the independent infinitesimal volume property of the Poisson process means this joint probability is

$$\text{Probability}[z_1 \in (v_1 + \Delta v), z_2 \in (v_2 + \Delta v), \ldots z_k \in (v_k + \Delta v)]$$

$$= \prod_{i=1}^{k} \left[\int_{v_i}^{v_i + \Delta v} m(v) dv \right] \left[e^{-\int_{V-k\Delta v} m(v) dv} \right] \tag{A.3}$$

where $m(v)$ is the count intensity of the received field over the area A during the time T, and indicates the rate of count occurrences. Now as $\Delta v \to 0$, Eq. (A.3) approaches

$$\prod_{i=1}^{k} m(v_i)\Delta v \left[e^{-\int\limits_{V} m(v)\,dv} \right] \tag{A.4}$$

which is the basic Poisson assumption. On the other hand, for small Δv, Eq. (A.3) could have been written as

$$\text{Probability}[z_1 \in (v_1 + \Delta v), z_2 \in (v_2 + \Delta v), \dots z_k \in (v_k + \Delta v)]$$
$$= k! \text{Prob}[z_1 = v_1, \dots z_k = v_k]\,(\Delta v)^k \tag{A.5}$$

Where the $k!$ term takes into account all the ways in which k events can occur in k infinitesimal volumes. Equating Eqs. (A.4) and (A.5) identifies the joint density as

$$p_z(z_1, z_2, \dots z_k, k) = \frac{1}{k!} \prod_{i=1}^{k} m(z_i) \left[e^{-\int\limits_{V} m(v)\,dv} \right] \qquad z_i \in V \tag{A.6}$$

Since the conditional density $p_z(z_1, z_2, \dots z_k/k)$ is obtained as

$$p_z(z_1, z_2, \dots z_k/k) \equiv \frac{p_z(z_1, z_2, \dots z_k, k)}{p_k(k)} \tag{A.7}$$

dividing Eq. (A.6) by (A.1) yields for $p_z(z_1, z_2, \dots z_k/k)$

$$p_z(z_1, z_2, \dots z_k/k) = \prod_{i=1}^{k} \frac{m(z_i)}{\left(\int\limits_{v} m(v)\,dv \right)^k} \tag{A.8}$$

Notice that the desired conditional density for the occurrence locations factors into the product of the individual densities

$$\frac{m(z)}{\int\limits_{V} m(v)\,dv} \tag{A.9}$$

for each i. This implies that Poisson occurrences are independent, each having a location density over V given by the normalized count intensity process $m(z)/m_v$. Notice also that this probability density is non-negative, integrates to unity, and is valid for any volume V. The time-space duality is portrayed in Figures A.1 and A.2, where we have used the normalization in Eq. (2.46).

$$n(t,r) \equiv \frac{\eta}{hf} I(t,r)$$ Number of electrons per unit area per unit time

$$\alpha = \frac{\eta}{hf}$$

$$n(t) \equiv \int_A n(t;r)dA$$

$$n(r) \equiv \int_t^{t+T} n(t;r)dt$$

$$p_T(t_j) = \frac{n(t_j)}{\int_t^{t+T} n(t)dt}$$

$$p_A(r_i) = \frac{n(r_i)}{\int_A n(r_i)dA}$$

Figure A.1 Key photo-count equations.

$$p_T(t_j) = \frac{n(t_j)}{\int_T n(t_j)dt} = \frac{\bar{n}\left[(1-m_t(t_j))\right]}{\bar{n}T} = \frac{\left[(1-m_t(t_j))\right]}{T}$$

$$p_A(r_i) = \frac{n(r_i)}{\int_A n(r_i)dA} = \frac{\bar{n}T[(1-m_r(r_i))]}{\bar{n}TA} = \frac{[(1-m_r(r_i))]}{A}$$

Normalized signal power becomes

$$\int_T m_t^2(t)dt \int_A m_r^2(r)dr = \bar{m}_t^2 T \bar{m}_r^2 A$$

Figure A.2 Identification of image in photo-count equations.

Appendix B: Propagation of finite beams in water

Finite beams

Consider the general equation of radiative transport

$$N(\theta, \vartheta, z) - N(\theta, \vartheta, z + \Delta z) = -K(\theta, \vartheta, z)N(\theta, \vartheta, z)\Delta z$$

with solution

$$\frac{\Delta N(\theta, \vartheta, z)}{N(\theta, \vartheta, z)} = -K(\theta, \vartheta, z)\Delta z$$

or

$$N(\theta, \vartheta, z) = N(\theta, \vartheta, z_0)e^{-\int_{z_0}^{z} K(\theta, \vartheta, z)\,dz}$$
$$= N(\theta, \vartheta, z_0)e^{-K(z - z_0)} \quad K \text{ constant}$$

This is the solution for the transfer of radiance in an infinite medium, as a function of z, since there is assumed to be no change in the radiance profile other than the translation in z.

Arnush and Heggestad [1,2], have introduced a solution to the radiative transport equation for small-angle scattering to be

$$f(\theta_r, \theta_\phi, r) \approx \frac{1}{(\pi U_0 R_1)^2} e^{\left[-az - \frac{(\theta_r - \theta_m)^2 + \theta_\phi^2}{U_\phi^2} - \frac{r^2}{R_1^2}\right]}$$

where

$$U_\phi^2 \approx \frac{sz^3 \bar{\theta}_{rms}^2}{2}$$

$$R_1^2 \approx \frac{2sz^3 \bar{\theta}_{rms}^2}{3}$$

$$\theta_m \approx \frac{3r}{2z}$$

If we assume that the receiver will integrate all the radiation over the hemisphere, this reduces to

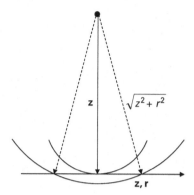

Figure B.1 Propagation geometry.

$$f(r) = \frac{1}{(\pi_0 R_1)^2} e^{\left[-az-\frac{r^2}{R_1^2}\right]}$$

We see in Figure B.1 this solution mapped on a sphere. In order to map it on a plane at a depth z, Karp derived an equation for the contribution to a point at $f(0,z)$, given the distribution over the surface $f(r,0)$[3]. This takes the form

$$f(0,z) = 2f(r,0) \int_0^{r_0} \frac{e^{-\left[a(z^2+r^2)^{1/2}\right] + \frac{r^2}{(2/3)s\bar{\theta}_{rms}^2(z^2+r^2)^{3/2}}}}{(2/3)s\bar{\theta}_{rms}^2(z^2+r^2)^{3/2}} r\, dr$$

Here $\bar{\theta}_{rms}^2 = \Omega_{rms}$, the rms scattering angle (radians or steradians) of the phase function. This equation treats $f(r)$ as a spatial impulse response, and integrates the contribution to $f(0,z)$ over a finite surface. It has been shown empirically that this converges to $f(0,0)e^{-kz}$ for $r_0 \to \infty$ (Figure B.2). If a circular spot of light, having radius r_s and one joule of energy, illuminates the surface of the ocean, the radial profile x_0 at a depth z can be written as,

$$\text{Loss} = \frac{\int_{-r_s}^{r_s} \int_0^{\sqrt{r_s^2-x^2}} \frac{e^{-a\left[z^2+y^2+(x-x_0)^2\right]^{1/2} - \frac{y^2+(x-x_0)^2}{2/3\, b\Omega_{rms}\left[z^2+y^2+(x-x_0)^2\right]^{3/2}}}}{2/3\, b\Omega_{rms}\left[z^2+y^2+(x-x_0)^2\right]^{3/2}} dy\, dx}{\pi r_s^2}$$

as shown in Figure B.3. Notice that to get the most uniform value, within 3 dB, for the depth of 100 meters, required a spot approximately 100 meters at the surface. Notice also that all the curves asymptote to the same e^{-kz} slope, the actual value depends upon the radius of the spot on the surface. This is shown again in Figure B.4 where we show the evolution of radial profile of a spot 128 meters on the surface observed at various depths.

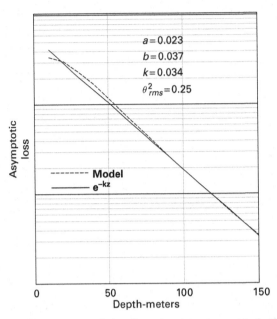

Figure B.2 Asymptotic irradiance loss in water versus depth.

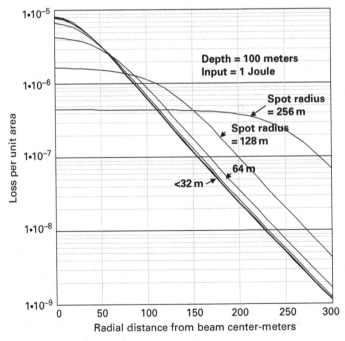

Figure B.3 Loss per unit area versus radial distance from beam center for point illumination.

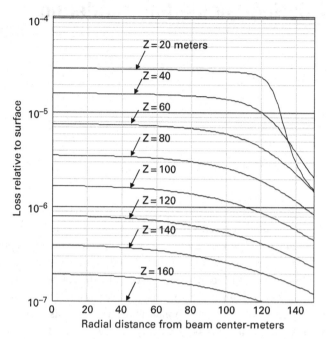

Figure B.4 Loss relative to surface versus radial distance from beam center for 120 meter beam radiance.

References

1. D. Arnush. Underwater light-beam propagation in the small angle scattering approximation. *Journal of the Optical Society of America*, **62** (1972), p. 1109.
2. H. M. Heggestad. Optical communications through multiple scattering media. MIT, RLE Technical Report 472, November 1968.
3. S. Karp. Optical communications between underwater and above surface (satellite) terminals. *IEEE Transactions on Communications*, **24**(1) (1976), pp. 66–81.

Appendix C: Non-Lambertian scattering

An excellent characterization of reflecting surfaces has been given by Ruze [1]. Although it was intended for coherent antenna structures, it is directly applicable to optical surfaces. As a model he uses a two-level random process, one with a short coherence length which he calls roughness, and a second with a large coherence length, c, which he calls smoothness (Figure C.1). In Chapter 2 we describe the texture as being rough and inhomogeneous, which corresponds to what Ruze has called roughness. However, what Ruze calls smoothness is usually referred to as flatness for optical surfaces. For a surface to be of "optical quality", both the smoothness and roughness should be less than $\lambda/20$.

For the purposes of lidar, we see that the cross section of a surface can have one of three characteristics. Thus consider the cross section of a corner cube described in Chapter 9, example 9.2. The cross section of an "optical quality" corner cube with a side of length L is

$$\sigma_L = \frac{4\pi(0.579)L^2}{\lambda^2} \qquad\qquad \text{C.1}$$

If the surface is "smooth" to a value $c < L$, then the cross section is reduced to

$$\sigma_c = \frac{4\pi(0.579)c^2}{\lambda^2}. \qquad\qquad \text{C.2}$$

This reflector returns the incident radiation into a solid angle equal to $\Omega_c = \lambda^2(\sigma_c)$.

Finally, if the surface is rough, hence is Lambertian, the cross section becomes

$$\sigma_r = 3x\frac{(0.579)L^2}{\pi} \qquad\qquad \text{C.3}$$

because the return, being incoherent, will add from each third of the corner cube.

We point out that a commercial product called "Scotchlite" is a tape with tiny crystals imbedded into it. The tapes come with miniature corner cubes (crystals) with one of two dimensions, 4 mil and 17 mil. A mil is 1/1000th of an inch or 25 μm. These are marketed as highly retro-reflecting tapes.

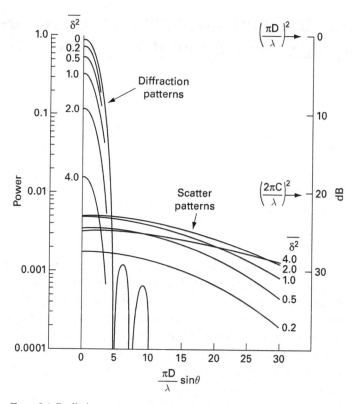

Figure C.1 Radiation patterns of phase distored circular aperture, 12 dB illumination taper, $D=20c$.

References

1. J. Ruze. Antenna tolerance theory – a review. *Proceedings of the IEEE*, **54**(4) (1966), pp. 633–641.

Appendix D: Communications noise sources besides signal/background shot noise

In Chapter 6, we considered signal and background shot noises as the limiting factor in the signal-to-noise ratio determination. Although in many cases this is accurate, there are a number of other noise sources that will plague the optical systems engineer. These include thermal noise, flicker noise and dark current. In this appendix, we will review the various noise sources we might need to consider in a communication receiver.

Thermal noise, or Johnson noise, is created by the thermal fluctuation of electrons in the electronics of an optical receiver [1, pp. 145–148]. Given a detector bandwidth B, the thermal noise power is given by expression.

$$P_{TH} = 4kTB \tag{D.1}$$

where

$k \equiv$ Boltzmann's constant

$$= 1.380\ 6504(24) \times 10^{-23}\,\text{J/K} \tag{D.2}$$

$T \equiv$ receiver temperature in degrees Kelvin (K)

$B \equiv$ detector bandwidth in hertz (Hz) or $(\text{seconds})^{-1}$

This follows the derivation in Chapter 1. Equation (D.1) is the same form for thermal noise one has with radio frequency systems, which is not too surprising.

In any photo-emissive or photo-voltaic detector, a dark current is created in the absence of any external optical source and this in turn creates shot noise within the receiver [1, pp. 148–150]. This is a time-dependent Poisson process similar to what was outlined in Chapter 3. Specifically, the probability that the number of dark current electrons emitted in a time period τ is exactly the integer k is given by

$$P(U_{dc,\tau} = k) = \frac{(\mu_{dc,\tau})^k\, e^{-\mu_{dc,\tau}}}{k!}$$

where $\mu_{dc,\tau} \equiv \tau\, I_{dc}/q$ is the average number of dark current electrons released by the detector in a time period τ. Here I_{dc} is the dark current and q is the charge of an electron, which equals $1.60217646 \times 10^{-19}$ coulombs. It can be shown that the noise power spectrum density for this noise component is given by

$$G_{Idc} = G^2 q I_{dc} + G^2 I_{dc}^2 \delta(f) \tag{D.3}$$

with G being the post-detector gain current. This equation implies the noise power spectrum is composed of a flat spectrum defined by the first part of the equation and a direct current component by the second part of the equation. The total dark current shot noise power at a load resistor, R_L, created by the simple resistor-capacitor filter of bandwidth B can then be written as

$$P_{dc} = 2G^2 q I_{dc} B R_L \tag{D.4}$$

The fluctuations in the emissions from spots on a vacuum tube cathode create flicker noise [1, p. 152]. The spectrum of this noise is inversely proportional to frequency for frequencies down to 1 Hz and to the square of the photocathode current, I_k. Its power spectrum density is given by

$$G_{IF} = \alpha_F G^2 I_k^2 / f \tag{D.5}$$

where α_F is a proportionality constant. The expression cannot hold down to zero frequency since the noise power must be bounded. Flicker noise is associated mainly with thermionic emission, so it is not a significant noise factor in photo-emissive tubes.

Semiconductor detectors carrying a steady current exhibit a noise effect called current of $1/f$ noise (current noise), which is a one-sided power spectrum inversely proportional to frequency to below 1 Hz [1, pp. 152–153]. The noise spectrum depends upon the square of the post-detector amplified current and the operational environment – most notably upon the humidity, but not upon the temperature of the device. The physical mechanism producing current noise is believed to be the trapping of charge carriers near the surface of a material. The power spectrum density of current noise is written as

$$G_{IC}(f) = \alpha_C G^2 I_P^2 / f \tag{D.6}$$

where α_C is a proportionality constant and I_P is the steady current in the semi-conductor detector.

In a heterodyne, or homodyne, optical receiver in which optical mixing occurs, the spectral (line) shape of the transmitter and local oscillator lines becomes significant because the laser lines are essentially shifted intact to a lower radio frequency called an intermediate frequency (IF) [1, pp. 153–155]. Phase noise arises from the random frequency shifting of the carrier and local oscillator lasers as a result of spontaneous emission quantum noise and environmental vibrations of the laser cavity mirrors. The quantum noise is a fundamental irreducible cause of frequency fluctuations. In most applications, the quantum noise is hidden by the much stronger environmental noise that comes from turbulent atmospheric signal fading. Carrier frequency fluctuations due to atmospheric turbulence are another major cause of phase noise.

Relative intensity noise (RIN) describes the instability in the power level of a laser. For example, RIN can be generated from cavity vibration, fluctuations in the laser gain medium or simply from transferred intensity noise from a pump source. Since intensity noise typically is proportional to the intensity, the relative intensity noise is typically independent of laser power. RIN typically falls off with frequency and is a kind of pink noise. It is typically measured by sampling the output current of a photo-detector over time and transforming this data set into frequency with a fast Fourier transform. For a system without optical preamplification, say with an avalanche photo diode (APD), the RIN noise power can be written as

$$P_{RIN} \propto \left(\frac{q\eta}{hf}\right)^2 (RIN) \, P_T \, B \, M^n \tag{D.7}$$

where

P_T = transmitter optical power (W)
q = electron charge
h = Planck's constant = 6.626×10^{-34} joule-seconds
f = frequency of light (Hz)
η = detector quantum efficiency (amps /watts)
M = APD gain.

Many optical communications systems use an optical fiber amplifier to amplify the received signal. The noise analysis in this case is a little more complicated as it involves spontaneous and stimulated emissions in the fiber amplifier [2]. Specifically, signal-spontaneous beat noise is proportional to

$$2GI_s I_{sp} B_e / B_o \tag{D.8}$$

and the spontaneous-spontaneous beat noise is proportional to

$$\frac{1}{2} I_{sp}{}^2 [B_e(2B_o - B_e)/B_o{}^2] \tag{D.9}$$

where
B_e ≡ bandwidth of the electrical filter used in the electrical receiver circuit
G ≡ amplifier gain
I_s ≡ photocurrent generated by received bit "1" signal in the absence of a fiber amplifier

$$= (q\eta/hf)P_s \tag{D.10}$$

P_s ≡ receiver bit "1" optical power before the amplifier
I_{sp} ≡ photocurrent generated by the spontaneous emissions (including both polarizations) at the output of the detector

$$= 2n_{sp}(G-1)eB_0 \tag{D.11}$$

n_{sp} ≡ inversion parameter
B_o ≡ optical passband bandwidth.

References

1. W. K. Pratt. *Laser Communications Systems*. John Wiley & Sons, New York (1969).
2. P. Becker, N. Olsson and J. Simpson. *Erbium-Doped Fiber Amplifiers Fundamentals and Technology*. Academic Press, New York (1999).

Index

Printed in the United States
By Bookmasters